SpringerBriefs in Materials

Series Editors

Sujata K. Bhatia, University of Delaware, Newark, USA

Alain Diebold, Schenectady, USA

Juejun Hu, Department of Materials Science and Engineering, Massachusetts Institute of Technology, Cambridge, USA

Kannan M. Krishnan, University of Washington, Seattle, USA

Dario Narducci, Department of Materials Science, University of Milano Bicocca, Milano, Italy

Suprakas Sinha Ray , Centre for Nanostructures Materials, Council for Scientific and Industrial Research, Brummeria, Pretoria, South Africa

Gerhard Wilde, Altenberge, Germany

The SpringerBriefs Series in Materials presents highly relevant, concise monographs on a wide range of topics covering fundamental advances and new applications in the field. Areas of interest include topical information on innovative, structural and functional materials and composites as well as fundamental principles, physical properties, materials theory and design. **Indexed in Scopus (2022).**

SpringerBriefs present succinct summaries of cutting-edge research and practical applications across a wide spectrum of fields. Featuring compact volumes of 50 to 125 pages, the series covers a range of content from professional to academic. Typical topics might include

- A timely report of state-of-the art analytical techniques
- A bridge between new research results, as published in journal articles, and a contextual literature review
- A snapshot of a hot or emerging topic
- An in-depth case study or clinical example
- A presentation of core concepts that students must understand in order to make independent contributions

Briefs are characterized by fast, global electronic dissemination, standard publishing contracts, standardized manuscript preparation and formatting guidelines, and expedited production schedules.

Zakariya Zubair • Khubab Shaker • Asif Hafeez

Functional Polyurethane Coatings

Formulations, Properties, and Applications

 Springer

Zakariya Zubair
Department of Materials
National Textile University
Faisalabad, Pakistan

Khubab Shaker
Department of Textile Engineering
National Textile University
Faisalabad, Pakistan

Asif Hafeez
Department of Materials
National Textile University
Faisalabad, Pakistan

ISSN 2192-1091   ISSN 2192-1105   (electronic)
SpringerBriefs in Materials
ISBN 978-3-032-12729-7   ISBN 978-3-032-12730-3   (eBook)
https://doi.org/10.1007/978-3-032-12730-3

This Springer imprint is published by the registered company Springer Nature Switzerland AG
The registered company address is: Gewerbestrasse 11, 6330 Cham, Switzerland

If disposing of this product, please recycle the paper.

Preface

The research in field of coatings is driven by ever-increasing demands for superior performance, sustainability, and specialized functionalities across nearly every industrial sector. Few materials have demonstrated the versatility and capacity for innovation in this space as profoundly as polyurethane. With its inherent adaptability, polyurethane (PU) has long been a cornerstone of high-performance coatings. From providing exceptional abrasion and corrosion resistance in heavy-duty industrial environments to delivering high-gloss, aesthetically pleasing finishes in automotive and architectural applications, PU coatings have proven their value.

This book, *Functional Polyurethane Coatings: Formulations, Properties, and Applications*, arrives at a crucial time to address this exciting evolution. It serves as an essential guide that moves beyond the fundamental chemistry to focus on the cutting-edge of material design and engineering. The authors have meticulously compiled the latest research and industrial practices, offering a comprehensive and practical resource for both novices and seasoned professionals.

Readers will find a structured journey from the intricacies of modern PU formulation—including waterborne, high-solids, and non-isocyanate chemistries—to the detailed characterization of advanced properties. Crucially, the book closes the loop by providing an in-depth survey of the diverse applications where these functional coatings are indispensable, such as in aerospace, medical devices, advanced electronics, and sustainable construction. We are confident that the book will inspire further innovation in the dynamic and vital world of functional coatings.

Faisalabad-Pakistan
Faisalabad-Pakistan
Faisalabad-Pakistan

Zakariya Zubair
Khubab Shaker
Asif Hafeez

Contents

About the Authors

Zakariya Zubair obtained his doctorate in Mechanical Engineering from ENSISA, France in 2020, and is currently serving at the National Textile University at Faisalabad, Pakistan as Assistant Professor. He has 15 impact factor publications, five book chapters, and more than four conference talks to his credit. His areas of research include shape memory polymer composites, smart materials, and functional coatings.

Khubab Shaker is a highly accomplished researcher and academic in the field of textiles and composite materials. He completed his doctoral studies at the National Textile University in Faisalabad, Pakistan, in 2018. With an impressive track record, Dr. Shaker was included among the Top 2% cited scientist by Stanford University in 2024. Throughout his career, Dr. Shaker has made significant contributions to the field, with an extensive publication record. He has authored 97 research papers with an impact factor of 280 that have garnered substantial recognition within the academic community. In addition, he has edited eight books and contributed 29 chapters to various book publications. His expertise is further demonstrated through his participation in over 35 conference papers and keynote talks. Currently, Dr. Shaker holds the position of Assistant Professor at the Department of Textile Engineering at the National Textile University.

Asif Hafeez is currently an Assistant Professor at the Department of Materials in the National Textile University, where he has been actively involved in teaching and research. He completed his PhD in 2021 from the Universiti Teknologi Malaysia (UTM), Malaysia and MSc from KTH Royal Institute of Technology, Sweden. His research focuses on polyurethane coatings and adhesives, lamellar membranes, and membrane development for water purification. He has edited two books and published various book chapters, as well as research papers in well-renowned journals.

Chapter 1
Basics of Polyurethane Coating

1.1 Introduction

Polyurethane-based coatings represent a class of high-performance protective and decorative materials extensively utilized across various industrial sectors. Their widespread adoption is attributed to their exceptional versatility, mechanical durability, and resistance to environmental degradation. These coatings are synthesized through the polyaddition reaction between polyols and isocyanates, resulting in a crosslinked polymeric network capable of strong adhesion to a broad range of substrates, including plastics, wood, metals, and concrete. The performance attributes of polyurethane coatings include excellent corrosion resistance, hydrophobicity, resistance to ultraviolet (UV) radiation, moisture impermeability, protection against biological deterioration (e.g., fungal or microbial attack), and outstanding abrasion resistance. Additionally, their superior aesthetic qualities, such as gloss retention and surface smoothness, make them highly suitable for demanding applications in the automotive, industrial, and architectural sectors [1]. Polyurethanes (PUs) were first discovered in 1937 in Germany by 'Otto Bayer' and his colleagues at I.G. Farbenindustrie. Their work emerged as a significant development in polymer science, parallel to the research being conducted by 'Wallace Carothers' on polyamides (nylon) at DuPont. While Carothers focused on condensation polymers, Bayer's team pioneered the synthesis of polyurethanes through step-growth polymerization involving isocyanates and polyols, marking a foundational advancement in the field of polymeric materials [2].

Initially, Otto Bayer synthesized hydrophilic and infusible polyurea through the reaction of an aliphatic diisocyanate with a diamine. Building upon this early work, he later discovered that reacting an aliphatic isocyanate with a glycol yielded a new class of materials polyurethanes with highly desirable elastic properties. Owing to these properties, companies such as DuPont and Imperial Chemical Industries (ICI) commenced industrial production of polyurethanes in the early 1940s. However, the

Z. Zubair et al., *Functional Polyurethane Coatings*, SpringerBriefs in Materials, https://doi.org/10.1007/978-3-032-12730-3_1

outbreak of World War II led to a significant decline in the commercial growth of polyurethane materials. In response to the need for improved elastomeric performance, Bayer developed polyisocyanates in 1952, most notably toluene diisocyanate (TDI), which exhibited enhanced flexibility and durability.

In 1958, Schollenberger of B.F. Goodrich developed a novel, virtually crosslinked thermoplastic polyurethane (TPU) elastomer, marking a significant milestone in polyurethane technology. Around the same time, DuPont introduced a polyurethane-based spandex fiber known as Lycra, synthesized from polytetramethylene glycol (PTMG), 4,4′-diphenylmethane diisocyanate (MDI), and ethylene diamine. Subsequent developments in the polyurethane industry led to the commercialization of various specialized polyurethane products. These included Estane (B.F. Goodrich), Texin (Mobay), Pellethane (Upjohn), Desmopan (Bayer), and Elastollan (Elastogran), each tailored for specific mechanical, chemical, or processing performance characteristics [3].

The development of low-cost polyether polyols significantly enhanced the economic viability of polyurethane (PU) coatings, making them an attractive option for the automotive industry. Continuous advancements in processing and formulation technologies have further improved the performance characteristics of PU-based coatings. Today, these coatings represent one of the most widely utilized systems in both the construction and automotive sectors, owing to their superior color retention, high gloss, scratch resistance, and protective properties. Their combination of aesthetic performance and cost-effectiveness continues to drive their widespread adoption across diverse industrial applications.

1.1.1 Definition of Polyurethane Coatings

Polyurethane (PU) coatings are synthesized through the reaction between polyols and isocyanates ($-N=C=O$), resulting in the formation of urethane linkages and the development of a crosslinked polymeric network. Upon polymerization, these networks yield PU coatings characterized by exceptional mechanical strength, chemical resistance, and tailored performance properties. The reaction between isocyanates and polyols produces a continuous, flexible yet durable and rigid film, enabling PU coatings to serve both aesthetic and protective functions.

The versatility of PU-based coatings is achieved through precise control of the curing parameters and the chemical composition of the precursor materials. Curing can be initiated via thermal activation, ambient moisture, or by mixing the resin with a hardener in two-component systems. During curing, the liquid PU transforms into a solid, adherent film on the substrate surface. The resultant coating exhibits an optimal balance of flexibility, toughness, and resistance to environmental stressors such as moisture ingress, ultraviolet (UV) radiation, and aggressive chemicals, making PU coatings a preferred choice across industries where both high-quality aesthetics and superior surface protection are essential [1].

1.2 Composition of Polyurethane Coatings

Polyurethane (PU) coatings are multi-component systems composed of polyurethane precursors primarily polyols and isocyanates along with solvents, pigments or dyes, catalysts, and performance-enhancing additives. The formulation process involves mixing polyols and isocyanates in a defined molar ratio, in the presence of catalysts and other functional additives, initiating the curing reaction. The relative proportions of polyol and isocyanate significantly influence the final properties of the coating. Polyols impart flexibility and resilience to the PU matrix, whereas isocyanates function as crosslinking agents, contributing to structural stability and mechanical strength. Consequently, higher polyol content yields a more flexible coating film, while increased isocyanate content produces a more rigid and dimensionally stable film [4]. An exothermic polymerization reaction occurs between the isocyanate and polyol, resulting in the formation of polyurethane, as illustrated in Fig. 1.1.

1.3 Types of Polyurethane Coatings: Solvent-Based, Water-Based, and High Solids

Due to the versatile nature of polyurethane, various coating formulations can be designed to meet specific application requirements, including water-based, solvent-based, and high solid polyurethane systems. Each formulation is optimized for particular performance criteria and environmental conditions.

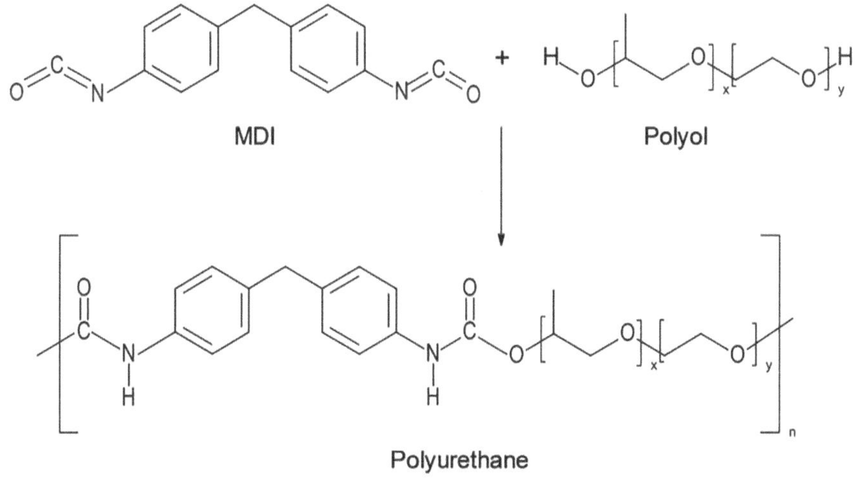

Fig. 1.1 Polyurethane manufacturing polymer reaction [5]

1.3.1   Solvent-Based Polyurethane Coatings

1.3.1.1   Composition and Properties

Solvent-based polyurethane (PU) coatings are formulated using organic solvents—such as toluene, xylene, ethyl acetate, and n-butyl acetate—as the primary medium in which the polyurethane resin is dissolved. These coatings are typically produced through the polymerization reaction between polyols and polyisocyanates, such as diphenylmethane diisocyanate (MDI) or toluene diisocyanate (TDI), resulting in the formation of polyurethane. In solvent-based PU coatings, the organic solvents play an integral role in adjusting the viscosity to facilitate smooth application onto the substrate. Following application, the solvents evaporate, and the coating undergoes curing through chemical reactions with polyol-based curing agents. This process yields a smooth, hard, and durable PU film with excellent adhesion and surface finish [6]. Solvent-based PU coatings generally exhibit relatively long drying times, often requiring several hours to a few days for complete curing, depending on environmental factors such as ambient temperature and humidity. Post-application cleaning of equipment typically necessitates the use of organic solvents or mineral spirits. These coatings are characterized by their high gloss, superior durability, and excellent resistance to heat, chemicals, and abrasion. Compared to water-based PU systems, solvent-based coatings offer higher coverage rates and produce a smoother, more leveled film with superior surface finish quality [7].

1.3.1.2   Applications

Due to their durability, aesthetic appeal, and resistance to harsh environmental conditions, solvent-based polyurethane (PU) coatings find extensive use across various sectors:

- **Wood products finishing**: Wooden furniture, in particular, benefits from solvent-based polyurethane coatings, which impart a smooth, high-gloss finish that enhances the natural grain and aesthetic appeal of the wood. Their compatibility with a wide range of wood species makes them suitable for diverse furniture styles and design requirements.
- **Metallic surface coatings**: Solvent-based PU coatings exhibit excellent adhesion to metallic substrates while providing effective resistance against moisture, corrosion, and chemical attack. Consequently, they are extensively used for protecting metallic components exposed to aggressive conditions, particularly in marine environments.
- **Automotive industry**: Automotive finishes demand both durability as well as high gloss appearances which is attributed to solvent-based polyurethane coatings. In addition, these coatings provide long-term protection against environmental stressors such as ultraviolet (UV) radiation and chemical exposure, which

can otherwise cause fading, chalking, and degradation of the underlying paint layer.

- **Industrial flooring**: Solvent-based polyurethane is widely used for hardwood flooring due to its excellent durability and ability to withstand heavy foot traffic. Aliphatic solvent-based PU coatings are also applied to concrete surfaces, offering superior abrasion resistance, chemical resistance, and long-term performance in demanding environments [7].

1.3.1.3   Advantages

- **Durability**: Solvent-based PU coatings form a dense, crosslinked protective layer on the substrate surface, exhibiting excellent resistance to heat, chemicals, and surface abrasion, including scratches.
- **Aesthetic appeal**: Highly glossy, smooth, and high coverage are achieved from solvent-based PU coatings.
- **Economical application**: Its excellent hiding power allows full substrate coverage with fewer coats, thereby reducing labor costs.
- **Ultra violet light resistance**: Solvent-based PU coatings exhibit excellent ultraviolet (UV) resistance, making them well-suited for use in high-profile outdoor structures exposed to prolonged sunlight [7].

1.3.1.4   Disadvantages

- **VOC emissions**: During application, solvent-based PU coatings release volatile organic compounds (VOCs), which pose significant health and environmental hazards. Therefore, the use of appropriate personal protective equipment (PPE) is essential during handling, and adequate ventilation must be ensured throughout the application process [8].
- **Longer drying times**: Drying of solvent-based PU coatings occurs upon complete evaporation of the solvents from the applied film, a process that may require several hours to several days depending on environmental conditions. Consequently, these coatings generally exhibit longer drying times compared to other PU coating systems.
- **Need of safe storage and disposal**: Solvent-based PU coatings present several safety hazards due to their high flammability and potential to cause skin irritation upon direct contact. Therefore, they must be stored in a cool, dry location away from heat sources, and any unused material should be disposed of at an authorized hazardous waste facility in accordance with safety regulations.
- **Proper ventilation**: Adequate ventilation must be ensured during the application of solvent-based polyurethane coatings to minimize inhalation exposure to harmful volatile organic compounds (VOCs) [7].

1.3.2 Water-Based Polyurethane Coatings

In recent years, the growing demand for environmentally sustainable materials has driven extensive research and development of water-based film and coating formulations. To address this need, waterborne polyurethanes (WBPUs) have been developed, which, during coating application or self-standing film formation, release only water into the atmosphere rather than volatile organic compounds (VOCs) typically emitted by conventional solvent-based polymers. Among their various applications, water-based paints and coatings remain the most widely studied. Advances in WBPU technology have been achieved through the strategic selection of hydroxyl-containing components, as well as the optimization of chain extenders and emulsifiers. Furthermore, to meet specific performance requirements, composite WBPU formulations have been enhanced through the incorporation of organic and inorganic micro- and nanoparticles [9].

1.3.2.1 Composition and Properties

Water-based PU coatings, also referred to as polyurethane dispersions (PUDs), utilize water or eco-solvents as the primary dispersion medium, thereby complying with environmental protection regulations and significantly reducing VOC emissions. Similar to conventional polyurethanes, WBPUs are composed of alternating structural domains: soft segments (SS), derived from polyols, which impart flexibility and elasticity, and hard segments (HS), formed by the reaction of isocyanates with low molecular weight chain extenders and, in many cases, internal emulsifiers, which provide rigidity, strength, and phase separation essential for film formation and performance [9]. The emulsified polyurethane used in water-based PU coatings is typically synthesized from a diisocyanate, a polyol, and a dispersing agent such as dimethylolpropionic acid (DMPA), with a diamine serving as the chain extender for the prepolymer. In water-based PU coatings, also known as polyurethane dispersions (PUDs), the polyurethane polymer exists as dispersed particles swollen with water. The resulting colloidal suspension must remain stable over time, ensuring that the dispersed polymer nanoparticles (discontinuous phase) do not sediment and that the properties of the continuous phase remain largely unchanged. Instability can occur when polymer nanoparticles diffuse and coalesce into larger aggregates, thereby reducing the total surface area and adversely affecting the coating's performance [10].

Several synthetic routes are employed for the preparation of waterborne polyurethane (WBPU) dispersions, including the acetone (or solution) process, the prepolymer mixing process, the hot-melt process, and the ketimine/ketazine process. In the prepolymer mixing process, an isocyanate-terminated prepolymer is synthesized from a polyol, a diisocyanate, and a dihydroxyalkanoic acid. This prepolymer is subsequently neutralized, dispersed in water, and chain-extended. In an alternative approach, the isocyanate-terminated prepolymer may be prepared either in bulk or with the aid of a temporary solvent or reactive diluent to reduce viscosity. This facilitates the dispersion of a low-molar-mass prepolymer in water while maintaining manageable viscosity during processing. The final molecular weight is achieved via chain extension within the dispersed polymer particles, typically through the reaction of a diamine with the isocyanate-terminated chains, introducing urea linkages in addition to the pre-existing urethane bonds formed during prepolymer synthesis [11].

The primary distinction between solvent-based PU and waterborne polyurethane (WPU) lies in the incorporation of hydrophilic segments into the polymer backbone. These segments are introduced through diols containing acid functionalities, such as carboxylic acid, sulfonate, or quaternary ammonium groups, which act as internal emulsifiers, enabling the polymer to disperse in water [12]. Figure 1.2 illustrates the complete chemical reaction pathway for the synthesis of water-based polyurethane dispersion via the prepolymer mixing process.

1.3.3   UV-Curable Water-Based PU Coatings

Waterborne polyurethane coatings prepared via UV-induced photo-polymerization (UV-WPU) have gained significant attention in response to increasingly stringent environmental regulations. Developed as sustainable alternatives to solvent-based polyurethanes, these coatings are widely applied to wood, paper, plastics, metals, and glass due to their excellent physicochemical, rheological, and optical properties. Over the years, extensive research has been conducted on various UV-WPU formulations, leading to notable advancements in performance and application versatility.

UV-WPU dispersions are commonly synthesized via the acetone method using a three-step process. First, an excess of difunctional diisocyanate is reacted with a polyol and end-capped with hydroxyl methacrylates in the presence of acetone. To enhance water dispersibility, part of the chain extender is substituted with a carboxylic acid-functionalized diol. In the second step, the carboxylic acid groups are neutralized, and the acrylic end-capped polyurethane prepolymer is dispersed in water. Finally, acetone is removed under vacuum, yielding a WPU dispersion suitable for subsequent free-radical UV curing. The resulting dispersions typically exhibit a pH of 7–8, a solid content of 30–40 wt.%, and stability over several months [14]. Figure 1.3 illustrates the curing steps of UV-curable water-based polyurethane coatings. In the first stage, water evaporates from the applied coating to form a dry film. This is followed by UV irradiation, which initiates crosslinking reactions, leading to complete curing of the film.

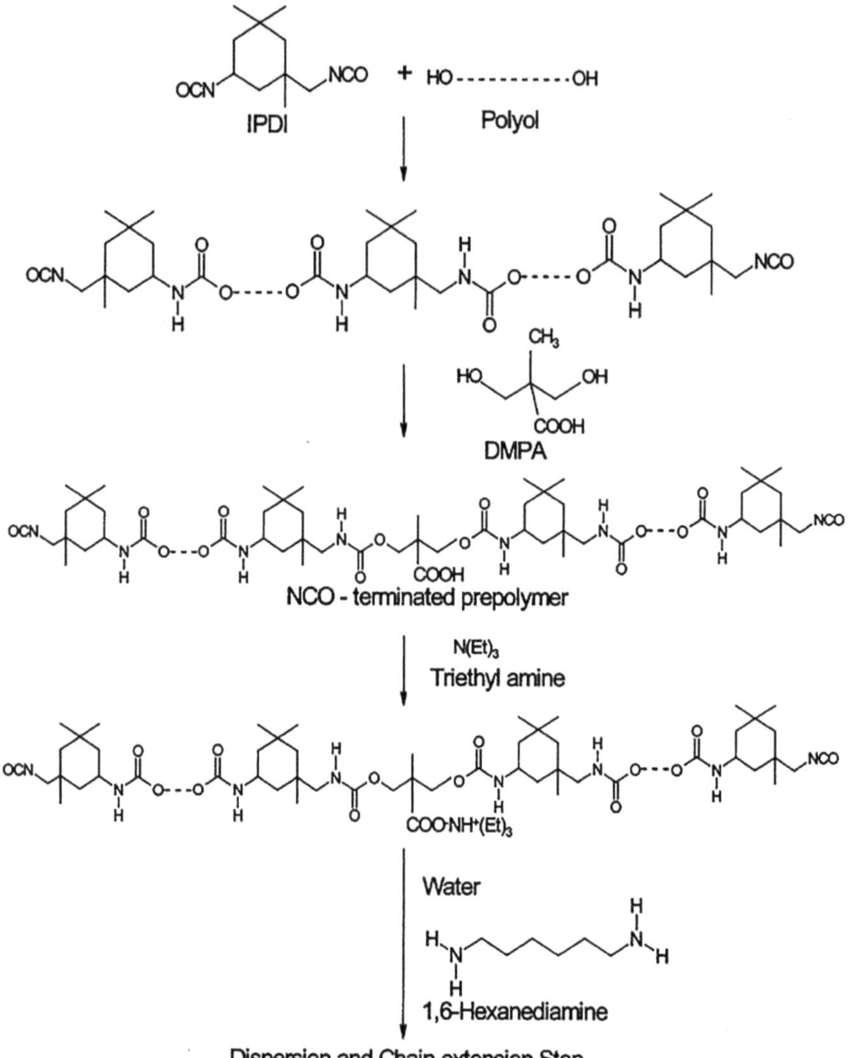

Fig. 1.2 Chemical reaction involved in manufacturing of water-based PU coating [13]

1.3.3.1   Applications

Water-based polyurethane (WBPU) coatings have found diverse applications owing to their environmentally friendly nature and favorable mechanical properties. Common application methods for WBPU systems include brushing, dipping, spraying, and rolling, with curing either physical or chemical sometimes required to complete film coalescence or achieve crosslinking. WBPU dispersions serve as versatile

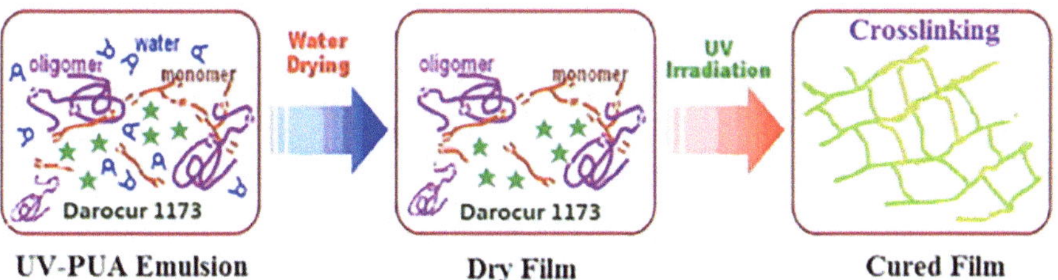

Fig. 1.3   Curing process of UV cures water-based PU coating [15]

platforms for formulating coatings with properties such as scratch resistance, durability, and excellent wetting and adhesion to various substrates. By incorporating specific pigments and additives, these dispersions can be tailored for use on wood, concrete, plastics, and metals. The properties of the final coating can be further tuned by varying the type of polyol used. Waterborne polyurethane (WBPU) coatings can also achieve improved corrosion protection through the incorporation of functional nanomaterials. For example, covalently and noncovalently grafted graphene oxide (GO) nanosheets have been used to enhance the barrier properties of WBPU coatings, reducing the penetration of water, oxygen, and corrosive ions. This approach significantly improves the long-term durability and corrosion resistance of coated substrates, making it suitable for applications in marine, industrial, and metallic structures [16]. For example, García-Pacios et al. compared dispersions synthesized from polyether, polyester, and polycarbonate diols and found that the polycarbonate diol-based system exhibited shorter drying times and superior adhesion on stainless steel [17]. Additionally, this particular coating presented higher chemical resistance to ethanol offers higher gloss and less yellow color.

Although WBPU systems generally exhibit lower performance compared to their solvent-based counterparts, they maintain a well-established presence in the coatings market. To address their inherent limitations, various strategies have been explored to enhance their properties and expand their applicability. For instance, soft-feel coatings for plastic substrates in automotive interiors—traditionally formulated using two-component solvent-based polyurethanes are now increasingly being replaced by WBPU formulations, offering a more environmentally friendly alternative without significantly compromising performance [18]. The incorporation of nanomaterials has emerged as an effective strategy to enhance or tailor the final properties of coatings. Furthermore, depending on the intended application, a range of standardized methods and testing protocols can be employed to characterize these materials and ensure compliance with relevant performance criteria [19].

1.3.3.2   Advantages

Low VOC emissions: One of the key advantages of waterborne polyurethanes (WBPU) is their significantly reduced emission of volatile organic compounds (VOCs). This feature, combined with excellent performance at low temperatures, makes them particularly attractive for flooring applications.

Odoreduction: Modern WBPU formulations contribute to a substantial reduction in odor during and after application.

Compatibility with sensitive substrates: WBPUs can be applied to surfaces that are sensitive to traditional solvent-based systems, such as certain plastics.

Lower toxicity and safer handling: They contain fewer toxic components, making them safer and more convenient to use compared to solvent-based alternatives.

Reduced aromatic solvent content: WBPUs have considerably lower levels of aromatic solvents, improving both environmental and occupational safety.

Effective crosslinking chemistry: WBPUs support highly efficient crosslinking reactions and can be used in polyurethane dispersions (PUDs), which is generally not feasible with conventional solution polymers.

Facilitated auto-oxidative crosslinking: Auto-oxidative crosslinking is more readily achieved in WBPUs. While possible in solvent-based systems, achieving similar results typically requires lower solids content and/or lower initial molecular weight. This enables WBPUs to attain desired hardness and resistance properties more quickly, even without accelerated curing methods such as applied heat making them especially suitable for high-traffic flooring applications [20].

Easy equipment cleaning: Application tools and equipment can be cleaned using only water and mild soap, eliminating the need for harsh chemical solvents [8].

1.3.3.3   Limitations

One disadvantage of WBPUs is that their hydrophilic groups can negatively impact surface properties, particularly water resistance, thereby reducing long-term performance. To address this, specifically modified polyols have been explored to enhance the final polymer properties. For instance, phosphorylated polyols have been used as starting materials, contributing to the formation of ionic soft segments that can either replace conventional emulsifiers during synthesis or introduce additional ionic centers. WBPUs synthesized from such polyols exhibit improved mechanical properties due to the increased hydroxyl functionality, which promotes higher crosslinking density. Furthermore, the inherent chemical structure of these polyols imparts enhanced anticorrosive behavior, thermal stability, and flame retardancy to the resulting PU films [20]. Effective surface cleaning or pretreatment is a critical prerequisite for achieving optimal adhesion of water-based PU coatings to the substrate [9].

1.3.4   High Solids Polyurethane Coatings

1.3.4.1   Composition and Properties

Polyurethanes are commonly synthesized by reacting a hydroxyl-functionalized monomer with an isocyanate-containing monomer, resulting in the formation of a urethane linkage. The versatility of polyurethane primarily depends on the nature

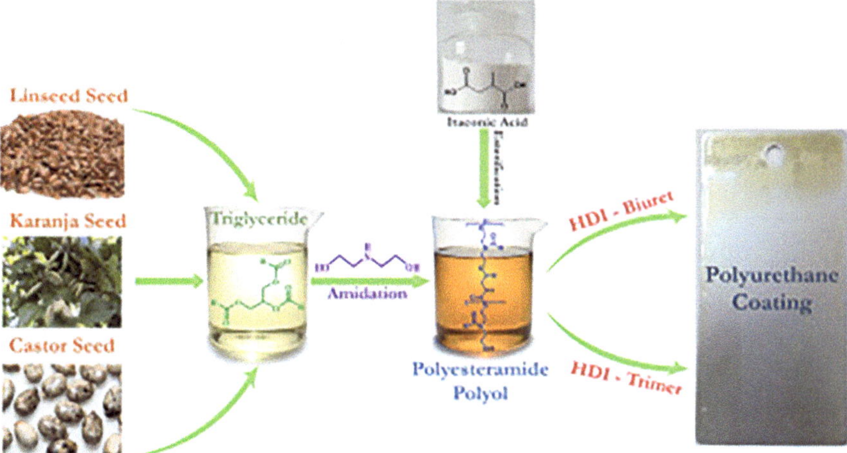

Fig. 1.4 Polyurethane coatings by using natural bio-based oil extracted from plants [25]

and molecular characteristics of its key reactants polyols and isocyanates. In PU coating production, the most common polyols are petrochemical-based polyester, polyether, and acrylic types. Polyester and acrylic polyols, under appropriate curing conditions, yield highly durable polyurethane films and are widely employed in high-performance coatings. In contrast, polyether polyols are typically utilized in highly flexible systems, such as sealants, interior applications, and primer formulations [21].

Traditionally, polyols used in the polyurethane industry have been derived from petrochemical sources. However, the growing global demand for crude oil coupled with its finite availability and volatile pricing has driven interest toward more sustainable and eco-friendly raw materials. This shift presents a significant opportunity for renewable feedstock-based polyols, such as those derived from vegetable oils, to gain a stronger foothold in the polyurethane market [22]. Conventional vegetable oil-based polyurethanes such as canola, sun flower, and flax as shown in Fig. 1.4 are typically synthesized by reacting diisocyanates with castor oil, a naturally occurring polyol that inherently contains hydroxyl groups [23]. Polyurethanes derived from castor oil find extensive applications in foams, interpenetrating polymer networks, and protective coatings [24]. Vegetable oil-based urethane varnishes as shown in Fig. 1.5, and urethane alkyds, typically prepared from diisocyanates, monoglycerides, and phthalic anhydride, have been in use for decades. In recent years, extensive research has focused on chemically modifying the double bonds in vegetable oils to produce multifunctional hydroxyl derivatives, which serve as renewable polyols for polyurethane synthesis [25]. High solids polyurethane (PU) coatings contain minimal to no solvents, with solids content ranging from 60% to 100% by weight. They employ low-viscosity resins that require no solvent dilution. Depending on formulation, these coatings are classified into one-component systems, which are moisture-cured, and two-component systems, which use blocked isocyanates to control

Fig. 1.5 Polyurethane coatings from vegetable oils and itaconic acid [27]

curing. High solids PU coatings produce thick films with excellent abrasion and impact resistance, while also offering superior elongation and flexibility [26].

1.3.4.2   Applications

High solids polyurethane coatings are applied in specific environments, including industrial facilities, marine structures, transportation equipment, and infrastructure exposed to harsh weather or chemical conditions:

- **Marine and industrial**: These coatings offer excellent corrosion resistance, making high solids PU coatings a preferred choice for marine applications and industrial equipment.
- **Automotive**: Owing to their high film-build capability and excellent mechanical properties, these coatings are widely used in automotive applications [26].
- **Pipelines**: These coatings offer a high stiffness-to-weight ratio, making them suitable for use in industrial pipelines [28].

1.3.4.3   Advantages

- **Low VOCs**: The minimal solvent content in high solids polyurethane coatings significantly reduces the emission of volatile organic compounds (VOCs), which are a major concern in environmental and occupational safety regulations. By keeping VOC levels low, these coatings meet or exceed strict environmental standards set by agencies such as the U.S. Environmental Protection Agency (EPA) and the European Union's REACH regulations. This not only helps industries avoid regulatory penalties but also contributes to improved air quality and safer working conditions for applicators.

- **Thick film buildup**: High solids polyurethane coatings form durable, seamless membranes that require fewer coats to achieve full protection and desired film thickness. Because these coatings have a high solids content (60–100%) and minimal solvents, each application deposits a thicker, denser layer of protective material compared to conventional solvent-based coatings. This means that the coating can achieve the required mechanical strength, chemical resistance, and barrier performance with fewer application passes, reducing labor time and overall project costs.
- **Versatility**: On flexible substrates such as rubber, flexible plastics, or fabric-reinforced materials the coating's inherent elasticity allows it to stretch and recover without cracking, ensuring long-term adhesion even under repeated bending or movement. On rigid substrates such as steel, aluminum, concrete, or hard plastics the coating's strong adhesion and high mechanical strength provide an excellent protective barrier against corrosion, abrasion, and environmental degradation.
- **Corrosion resistance**: In marine environments, they provide a strong barrier against saltwater corrosion, UV degradation, and biofouling, which helps extend the service life of ships, offshore platforms, and dockside structures. Their low permeability to water and chemicals ensures minimal degradation even under continuous moisture exposure. In industrial environments, high solids PU coatings resist abrasion, impacts, and exposure to oils, fuels, and a wide range of chemicals. This makes them effective for protecting equipment, storage tanks, pipelines, and heavy machinery operating under high stress or corrosive conditions [26].

1.3.4.4 Limitations

- **Shorter pot life**: In high solids coatings, the higher amount of functional groups can cause some drawbacks, such as reducing the pot life. This happens because the reaction rate at storage temperature increases with the concentration of functional groups, which shortens storage stability. Functional groups play an even more critical role in high solids coatings at least three times more than in conventional coatings with the same curing system since there is little to no solvent and higher functionality is needed to achieve proper chain extension [29].
- **Flocculation of pigment**: Preventing pigment flocculation is important for all types of coatings, but it becomes especially critical in high solids coatings. While conventional coatings rarely face major issues, in high solids systems flocculation can significantly increase viscosity before application. This happens because the continuous phase becomes trapped around the irregular surfaces of pigment aggregates, and these aggregates tend to crowd or pack together. A key factor in maintaining pigment stability is the thickness of the adsorbed layer on the pigment surface. However, the low molecular weight molecules used in resins for high solids coatings often cannot form a sufficiently thick adsorbed layer, making stabilization more challenging [29].

- **Cure window**: The curing window for high solids coatings is narrower than that of conventional coatings, making precise formulation and accurate control of baking temperature and time essential. In conventional coatings, small variations, such as ±10% in baking temperature, time, or catalyst amount have minimal impact because each resin molecule contains many hydroxyl groups, and even if 10% remain unreacted, the overall properties change only slightly. In contrast, high solids coatings are carefully optimized for functional group content to achieve the desired mechanical properties without increasing viscosity. As a result, even a small amount of unreacted functional groups can cause significant changes in the final coating performance [29, 30].
- **Sagging**: Sagging is more common in high solids coatings than in conventional ones because there is less solvent loss during application. For example, a high solids coating with 70–80% solids at the spray gun will still have about 75–80% solids once applied to the surface. In contrast, a conventional coating might start at only 20–30% solids at the gun but increase to 75–95% solids on the substrate as more solvent evaporates during application. This higher initial solids content in high solids coatings means the material is thicker and heavier on the surface, making it more prone to sagging [31, 32].

1.3.5   Comparison of Solvent, Water and High Solid PU Coatings

Polyurethane coatings whether water-based, solvent-based, or high solids—each have distinct advantages for specific applications. Solvent-based PU coatings provide excellent durability and attractive finishes but release high levels of VOCs, which harm the environment. In contrast, water-based PU coatings are eco-friendly, quick-drying, and easy to work with, though they have lower hiding power and often require more coats for complete coverage. High solids PU coatings combine strong performance with environmental benefits by minimizing solvent content. Emerging technologies, such as bio-based polyurethane dispersions (PUDs) and graphene-reinforced nanocomposites, are further enhancing the performance, versatility, and sustainability of PU coatings, keeping them highly relevant in modern industries [33].

1.3.6   Key Properties and Advantages of Polyurethane Coatings

1.3.6.1   Mechanical Strength and Durability

Polyurethane coatings are known for their high toughness, excellent tensile strength, and strong abrasion resistance. With tensile strengths typically ranging from 20 to 50 MPa, they are widely used in industrial, automotive, and flooring applications, where their exceptional wear resistance ensures long-lasting performance even under demanding conditions [3]. The abrasion resistance of PU coatings, evaluated using the Taber abrasion test (ASTM D4060), shows only 10–20 mg weight loss after 1000 testing cycles. In comparison, acrylic or alkyd-based coatings typically lose 50–100 mg under similar conditions. This demonstrates that PU coatings possess significantly higher abrasion resistance than acrylic and alkyd resins [3]. The high abrasion resistance of PU coatings indicates their ability to withstand greater mechanical stresses, thereby extending the service life of substrates by effectively protecting their surfaces. This performance is attributed to their unique molecular structure, which combines hard and soft segments. The hard segments, derived from isocyanates, provide strength and rigidity, while the soft segments, from polyols, offer flexibility and enhanced impact strength. This balanced combination allows PU coatings to absorb high levels of energy under impact without sustaining damage or cracking [34].

1.3.6.2   Chemical and Environmental Resistance

Polyurethane coatings exhibit excellent chemical resistance, maintaining stability when exposed to a wide range of chemicals such as solvents, acids, alkalis, fuels, and other corrosive agents. For instance, PU coatings remain intact even after prolonged exposure to 10% sodium hydroxide (alkali) or 10% sulfuric acid, whereas vinyl coatings and epoxies tend to degrade under these conditions. This resistance is primarily due to the dense polymer network formed through the crosslinking of isocyanates and polyols, as well as advanced formulations like hybrids or fluorination, which effectively limit chemical penetration into the coating structure.

In addition to chemical resistance, PU coatings offer strong weathering and UV resistance, ensuring long-term environmental durability. Accelerated weathering tests conducted in accordance with ASTM G154 show that aliphatic PU coatings maintain their gloss and color even after 2000 h of UV exposure, making them highly stable in sunlight. In contrast, aromatic PU coatings experience gloss loss and yellowing under similar conditions. Due to their superior UV stability, aliphatic PU coatings are ideal for outdoor applications such as aircraft exteriors, marine vessels, and architectural shading structures exposed to prolonged sunlight [35].

1.3.6.3   Adhesion and Flexibility

Polyurethane coatings can be applied to a wide variety of substrates due to their excellent adhesion properties. They are highly compatible with materials such as wood, concrete, and plastics, making them versatile for multiple industries. Adhesion performance is commonly evaluated using the pull-off test, as specified in ASTM D4541. Comparative testing of epoxy and PU-based coatings showed adhesion strengths of 10.48 MPa for epoxy and 8.60 MPa for polyurethane, confirming that while PU coatings have slightly lower adhesion than epoxy, they still provide strong and reliable bonding to diverse surfaces [36]. The polar nature of the polyurethane backbone enables it to bond effectively with substrates through hydrogen bonding and van der Waals forces. Surface preparation prior to coating such as proper sanding and cleaning—further enhances this adhesion, ensuring that PU coatings achieve strong bonding with the substrate. This improved adhesion contributes to the coatings' reliable performance, even in demanding applications and harsh environmental conditions [37].

Polyurethane coatings possess excellent flexibility and elasticity, primarily due to the polyol component in their structure. By varying the type and ratio of polyol, the flexibility of the coating can be tailored to meet specific requirements. Aliphatic-based PU coatings can exhibit elongation-to-break values exceeding 100%, allowing them to withstand significant deformation without cracking. This high flexibility makes them ideal for applications where structures experience thermal expansion, contraction, or mechanical bending. For example, PU coatings are commonly applied to car bumpers and panels, where they provide both durability and an attractive finish [38]. In polyurethane systems, various modifications can be introduced to further enhance flexibility. For instance, in one study, the isocyanate component was chemically modified, after which polyurethane polyrotaxane was synthesized using specific ratios of polyrotaxane and the modified polyisocyanate. This PU polyrotaxane was then dispersed in water to produce a highly flexible, waterborne PU coating. Such structural modifications improve chain mobility and elasticity, enabling the coating to withstand greater deformation without loss of performance [39].

1.3.6.4 Versatility in Formulation

Polyurethane (PU) coatings combine excellent mechanical strength, chemical resistance, strong adhesion to diverse substrates, and high flexibility. These properties can be tailored to meet specific application requirements by adjusting the isocyanate-to-polyol ratio, altering the type or structure of the isocyanate or polyol, incorporating functional additives, or modifying the synthesis process. Through such variations, PU coatings can be engineered to produce hard, elastomeric, soft, or high-gloss finishes. In response to environmental concerns and strict VOC regulations, waterborne PU coatings have gained significant popularity, offering strong performance while drastically lowering emissions producing up to nine times less VOCs than their solvent-based counterparts [40]. By incorporating various additives such as pigments, antimicrobial agents, and UV stabilizers, polyurethane (PU) coatings can achieve specific functionalities due to their excellent compatibility with these materials. Among the advanced PU coating types, self-healing PU coatings are highly demanded in the aerospace and automotive industries. These coatings can repair minor scratches through their dynamic bonding mechanisms, thereby enhancing durability, preserving surface aesthetics, and reducing maintenance needs [41]. This adaptability ensures that polyurethane coatings remain at the forefront of coating technology.

1.3.6.5 Thermal Stability

Polyurethane coatings typically show a thermal degradation onset temperature around 273 °C to 278 °C [42]. Polyurethane (PU) coatings can be further enhanced with additives, such as modified graphene oxide or carbon-based fillers, to significantly improve thermal stability enabling them to withstand temperatures up to 400 °C. This exceptional heat resistance makes PU coatings suitable for extreme environments, such as chemical reactors operating at elevated temperatures and aerospace applications where structures experience severe temperature fluctuations at high altitudes. The crosslinked molecular structure of polyurethanes further minimizes thermal degradation, ensuring consistent long-term performance under such demanding conditions [43].

1.3.6.6 Enhanced Durability and Longevity

The exceptional durability and long-lasting performance of polyurethane (PU) coatings result from their superior chemical resistance and mechanical strength. When compared to acrylic and epoxy coatings, PUs stand out as the most mechanically durable, offering outstanding flexibility, abrasion resistance, and toughness. These properties stem from strong urethane linkages and the ability to customize formulations with flexible polyols. Such characteristics make PUs ideal for flooring, automotive finishes, and wood coatings where scratch and impact resistance are crucial. However, in humid environments, their performance can be reduced by hydrolytic degradation unless stabilizers are incorporated. Epoxy coatings, although mechanically strong due to their highly crosslinked structure, are less flexible and more brittle than PUs. They deliver excellent hardness and adhesion but have lower impact and abrasion resistance, making them prone to cracking under mechanical stress. Their strength lies in corrosion-resistant applications, especially in static environments.

Acrylic coatings particularly thermoplastic types are the least mechanically durable, with lower toughness and abrasion resistance compared to PUs and epoxies. Thermosetting acrylics, with higher crosslinking, offer improved scratch resistance but still perform better in weathering and UV stability than in mechanical endurance, often requiring additives or protective overcoats for enhanced service life. Overall, PUs excel in flexibility and abrasion resistance, epoxies offer hardness and adhesion for static applications, and acrylics provide strong UV stability but lag in mechanical durability. This performance advantage makes PU coatings economically and operationally beneficial, particularly in harsh or high-wear environments [44].

1.3.6.7 Aesthetic Versatility

Polyurethane (PU) coatings offer exceptional aesthetic versatility, allowing finishes ranging from high-gloss to matte while maintaining excellent color retention. Their smooth, uniform surface enhances visual appeal, making them a popular choice in architectural, automotive, and consumer goods sectors. Aliphatic PUs, with stable urethane linkages, exhibit superior gloss retention and resistance to UV-induced degradation compared to aromatic PUs, which tend to yellow and lose gloss due to photodegradation of aromatic groups. The addition of UV absorbers and hindered amine light stabilizers (HALS) further improves gloss stability by preventing chain scission and surface damage under sunlight. In outdoor conditions, PU coatings generally retain gloss better than epoxies but may fall short of acrylics unless stabilized, positioning them as a strong option for applications where both appearance and durability are essential. Their ability to be tailored in color and texture adds to their appeal, for example, luxury automotive manufacturers use PU coatings to achieve deep, vibrant finishes that resist environmental wear while maintaining a showroom-quality look [45].

1.3.6.8 Environmental Adaptability

Advancements in polyurethane (PU) coating technology increasingly focus on environmental sustainability, with waterborne and high solids formulations achieving up to a 70% reduction in volatile organic compound (VOC) emissions compared to traditional solvent-based systems. These innovations align with stringent environmental regulations, including those established by the European Union and the U.S. Environmental Protection Agency, without compromising the mechanical, chemical, or aesthetic performance of the coatings. As a result, eco-friendly PU coatings are gaining widespread acceptance across industries, offering a sustainable alternative that meets both regulatory requirements and high-performance application demand [40]. Additionally, the durability of polyurethane (PU) coatings supports sustainability by extending service life and reducing the frequency of recoating, thereby conserving raw materials, lowering energy consumption, and minimizing waste generation. Their compatibility with recyclable substrates, including aluminum and certain thermoplastics, further strengthens their environmental profile by enabling end-of-life material recovery and reuse. This combination of long-term performance and recyclability positions PU coatings as a key material in sustainable manufacturing and lifecycle management strategies in high-performance applications.

Polyurethane (PU) coatings are considered more economical over their service life compared to epoxy- or alkyd-based coatings. Although their initial raw material and manufacturing costs are higher due to the use of expensive components [46], their extended service life and minimal maintenance requirements significantly reduce overall lifecycle costs. Studies have shown that industrial equipment coated with PU required 30% fewer repainting cycles, while marine applications saw a 40% reduction in recoating needs. Additionally, the superior protective performance of PU coatings allows a 50 µm thickness to deliver equivalent protection to a 100 µm epoxy coating. This reduced thickness requirement not only lowers raw material consumption but also minimizes application time and labor, making PU coatings a cost-effective choice in the long term [44].

References

1. Akindoyo JO, Beg MDH, Ghazali S, Islam MR, Jeyaratnam N, Yuvaraj AR (2016) Polyurethane types, synthesis and applications—a review. RSC Adv 6(115):114453–114482. https://doi.org/10.1039/C6RA14525F
2. Conti DAM, Nakamura DY (2024) Polyurethane-based coatings: innovations, properties, and applications. Eur J Emerg Phys Chem 1:01
3. Chattopadhyay DK, Raju KVSN (2007) Structural engineering of polyurethane coatings for high performance applications. Prog Polym Sci 32(3):352–418. https://doi.org/10.1016/j.progpolymsci.2006.05.003
4. Das A, Mahanwar P (2020) A brief discussion on advances in polyurethane applications. Adv Ind Eng Polym Res 3(3):93–101. https://doi.org/10.1016/j.aiepr.2020.07.002
5. Barszczewska-Rybarek I et al (2025) Characterization of changes in structural, physicochemical and mechanical properties of rigid polyurethane building insulation after thermal aging in air and seawater. Polymer Bulletin 79:1–23. https://doi.org/10.1007/s00289-021-03632-x
6. Dijkstra DJ, Brock T, Wilde H-W (2019) PU coatings. Polym Int 68(5):823–825. https://doi.org/10.1002/pi.5805
7. Kumar S. Solvent-based polyurethane resin. https://anrvrar.in/solvent-based-polyurethane-resin/. Accessed 11 Aug 2025
8. Luo Z, Shi Y, Zhao D, He M (2011) Synthesis of epoxidatied castor oil and its effect on the properties of waterborne polyurethane. Procedia Eng 18:31–36. https://doi.org/10.1016/j.proeng.2011.11.006
9. Mucci VL, Hormaiztegui MEV, Amalvy JI, Aranguren MI (2024) Formulation, structure and properties of waterborne polyurethane coatings: a brief review. J Adhes Sci Technol 38(4):489–516. https://doi.org/10.1080/01694243.2023.2240587
10. Rahman MM, Kim HD (2015) Waterborne polyurethane/oil fly ash composite: a new environmentally friendly coating material. J Adhes Sci Technol 29(24):2709–2718. https://doi.org/10.1080/01694243.2015.1087252
11. Tennebroek R et al (2019) Water-based polyurethane dispersions. Polym Int 68(5):832–842. https://doi.org/10.1002/pi.5627
12. Yu F et al (2016) Crosslinked waterborne polyurethane with high waterproof performance. Polym Chem 7(23):3913–3922. https://doi.org/10.1039/c6py00350h
13. Colloids and surfaces a: physicochem. ResearchGate. https://www.researchgate.net/publication/286005163_Colloids_and_Surfaces_A_Physicochem. Accessed 21 Aug 2025
14. Dall Agnol L, Dias FTG, Ornaghi HL, Sangermano M, Bianchi O (2021) UV-curable waterborne polyurethane coatings: a state-of-the-art and recent advances review. Prog Org Coat 154:106156. https://doi.org/10.1016/j.porgcoat.2021.106156
15. Xu H, Qiu F, Wang Y, Wu W, Yang D, Guo Q (2012) UV-curable waterborne polyurethane-acrylate: preparation, characterization and properties. Prog Org Coat 73(1):47–53. https://doi.org/10.1016/j.porgcoat.2011.08.019
16. Wen J-G et al (2019) Improvement of corrosion resistance of waterborne polyurethane coatings by covalent and noncovalent grafted graphene oxide nanosheets. ACS Omega 4(23):20265–20274. https://doi.org/10.1021/acsomega.9b02687
17. García-Pacios V. Incidence of the polyol nature in waterborne polyurethane dispersions on their performance as coatings on stainless steel. https://scholar.google.com/scholar_lookup?hl=en&volume=76&publication_year=2013&pages=1726-1729&journal=Prog+Org+Coat&issue=12&author=V+Garc%C3%ADa-Pacios&author=M+Colera&author=Y+Iwata&title=Incidence+of+the+polyol+nature+in+waterborne+polyurethane+dispersions+on+their+performance+as+coatings+on+stainless+steel&doi=10.1016%2Fj.porgcoat.2013.05.007. Accessed 4 Aug 2025
18. Zhou L, Koltisko B (2005) Development of soft feel coatings with waterborne polyurethanes. Jct Coatings Tech 2(15):54–60

19. Koleske JV. Paint and coating testing manual: of the Gardner-Sward handbook. ASTM International. https://scholar.google.com/scholar_lookup?hl=en&publication_year=2012&author=J.+Koleske&title=Paint+and+coating+testing+manual+-+15th+edition+of+the+Gardner-Sward+handbook. Accessed 4 Aug 2025

20. Wang S et al (2019) Synergetic enhancement of mechanical and fire-resistance performance of waterborne polyurethane by introducing two kinds of phosphorus–nitrogen flame retardant. J Colloid Interface Sci 537:197–205

21. Philipp C, Eschig S (2012) Waterborne polyurethane wood coatings based on rapeseed fatty acid methyl esters. Prog Org Coat 74:705–711. https://doi.org/10.1016/j.porgcoat.2011.09.028

22. Sharma V, Kundu PP (2006) Addition polymers from natural oils—a review. Prog Polym Sci 31(11):983–1008. https://doi.org/10.1016/j.progpolymsci.2006.09.003

23. Deka H, Karak N (2009) Bio-based hyperbranched polyurethanes for surface coating applications. Prog Org Coat 66(3):192–198. https://doi.org/10.1016/j.porgcoat.2009.07.005

24. Green chemistry and the biorefinery: a partnership for a sustainable future. https://www.researchgate.net/publication/221678971_Green_chemistry_and_the_biorefinery_A_partnership_for_a_sustainable_future. Accessed 12 Aug 2025

25. Kong X, Liu G, Qi H, Curtis JM (2013) Preparation and characterization of high-solid polyurethane coating systems based on vegetable oil derived polyols. Prog Org Coat 76(9):1151–1160. https://doi.org/10.1016/j.porgcoat.2013.03.019

26. Pélissier K, Thierry D (2020) Powder and high-solid coatings as anticorrosive solutions for marine and offshore applications? A review. Coatings 10(10):916. https://doi.org/10.3390/coatings10100916

27. Paraskar PM, Prabhudesai MS, Kulkarni RD (2020) Synthesis and characterizations of air-cured polyurethane coatings from vegetable oils and itaconic acid. React Funct Polym 156:104734. https://doi.org/10.1016/j.reactfunctpolym.2020.104734

28. Guan SW. 100% solids polyurethane and polyurea coatings technology

29. Hill LW, Wicks ZW (1982) Design considerations for high solids coatings. Prog Org Coat 10(1):55–89. https://doi.org/10.1016/0300-9440(82)80006-4

30. Bauer DR, Budde GF (1981) Crosslinking chemistry and network structure in high solids acrylic-melamine coatings. Indust Eng Chem Product Res Dev 20(4):674–679. https://doi.org/10.1021/i300004a016

31. Jones FN et al. Solventborne and high solids coatings. https://scholar.google.com/scholar_lookup?title=Solventborne+and+high+solids+coatings&author=Jones,+F.N.&author=Nichols,+M.E.&author=Pappas,+S.P.&publication_year=2017&pages=357%E2%80%93365. Accessed 12 Aug 2025

32. Wu S (1978) Rheology of high solid coatings. I. Analysis of sagging and slumping. J Appl Polym Sci 22(10):2769–2782. https://doi.org/10.1002/app.1978.070221005

33. Polyurethane paint & coatings: uses, chemistry, process & formulation. https://www.specialchem.com/coatings/guide/polyurethane-coatings. Accessed 11 Aug 2025

34. Ni H, Daum J, Soucek M, Simonsick W (2002) Cycloaliphatic polyester based high solids polyurethane coatings: I. The effect of difunctional alcohols. J Coatings Technol 74:49–56. https://doi.org/10.1007/BF02697983

35. https://content.e-bookshelf.de/media/reading/L-10210632-f0b1a9b6b0.pdf. Accessed 4 Aug 2025

36. Adhesion test ASTM D4541 results for PU coatings. https://scispace.com/search. Accessed 4 Aug 2025

37. Amrollahi M, Sadeghi GMM (2016) Assessment of adhesion and surface properties of polyurethane coatings based on non-polar and hydrophobic soft segment. Prog Org Coat 93:23–33. https://doi.org/10.1016/j.porgcoat.2015.12.001

38. Alfergani AA et al (2024) Polyurethane as a versatile polymer for coating and anti-corrosion applications: a review. Kompleksnoe Ispolz Min Syra 331(4):21–41. https://doi.org/10.31643/2024/6445.36

39. High-flexibility PU paint and preparing method thereof. https://scispace.com/papers/high-flexibility-pu-paint-and-preparing-method-thereof-2fik73b1t9. Accessed 4 Aug 2025
40. U. E. N. C. for E. Assessment. Waterborne polyurethanes and their properties. https://hero.epa.gov/hero/index.cfm/reference/details/reference_id/4810076. Accessed 4 Aug 2025
41. Wang H, Xu J, Du X, Du Z, Cheng X, Wang H (2021) A self-healing polyurethane-based composite coating with high strength and anti-corrosion properties for metal protection. Compos Part B Eng 225:109273. https://doi.org/10.1016/j.compositesb.2021.109273
42. Heat-resistance of bis(p-acetamidalphnoxy)dimethyl silane modified polyurethane coatings. https://scispace.com/papers/heat-resistance-of-bis-p-acetamidalphnoxy-dimethyl-silane-87nz29lblu0c. Accessed 4 Aug 2025
43. (2024) Thermal degradation of non-isocyanate polyurethanes. J Therm Anal Calorim. https://doi.org/10.1007/s10973-024-13306-1
44. https://content.e-bookshelf.de/media/reading/L-10210632-f0b1a9b6b0.pdf. Accessed 11 Aug 2025
45. Wicks ZW, Jones FN, Pappas SP, Wicks DA (2007) Organic coatings: science and technology, 3rd edn. Wiley-Interscience
46. Industries C. Polyurethane vs. epoxy. Copps Industries. https://www.coppsindustries.com/blog/polyurethane-vs-epoxy/. Accessed: 4 Aug 2025

Chapter 2
Formulation and Application Techniques

2.1 Raw Materials and Ingredients Used in Polyurethane Coating Formulations

The primary raw materials in polyurethane (PU) coatings are polyols and isocyanates, which react to form the urethane linkages that give the coating its characteristic properties. In addition to these two core components, various additives, diluents, crosslinkers, and catalysts are incorporated to tailor performance for specific application requirements. These supplementary materials can enhance properties such as curing speed, adhesion, flexibility, chemical resistance, weatherability, and surface appearance, enabling PU coatings to be optimized for diverse industries and environments.

2.1.1 Polyols

Polyols serve as the primary building block of the urethane linkage, supplying the hydroxyl groups that react with isocyanates to form polyurethane. They contain multiple hydroxyl groups in their molecular structure and can be derived from both petrochemical sources and natural oils. Typically, polyols have high boiling points and appear as viscous liquids. Based on their chemistry, polyols are classified into polyether polyols and polyester polyols, each imparting distinct properties to the resulting PU coatings.

Polyester polyols shown in Error! Reference source not found are produced through an esterification reaction between dicarboxylic acids (e.g., adipic acid) and polyhydric alcohols (e.g., glycols). PU coatings made with polyester polyols offer excellent adhesion, chemical resistance, and high toughness, making them suitable for demanding applications such as in the sports industry, where they can withstand

Z. Zubair et al., *Functional Polyurethane Coatings*, SpringerBriefs in Materials, https://doi.org/10.1007/978-3-032-12730-3_2

repeated impact and mechanical stress. Polyester polyols are generally synthesized from diols and dicarboxylic acids, giving them a robust backbone that contributes to the durability of the coating [1, 2] (Fig. 2.1).

Polyether polyols are produced through the etherification reaction of epoxides, such as ethylene oxide, with hydroxyl-containing compounds. Some common polyether polyols are shown in Fig. 2.2. PU coatings formulated with polyether polyols exhibit excellent moisture resistance, hydrolytic stability, and flexibility, which makes them ideal for applications in humid or water-exposed environments, such as marine equipment, outdoor structures, and waterproof coatings. The type of polyol (polyester vs. polyether) and the molecular weight of the polyol significantly influence the final coating's flexibility, hardness, and crosslinking density.

- Higher molecular weight polyols generally yield softer, more flexible coatings with lower crosslinking density.
- Lower molecular weight **polyols** tend to produce harder, more rigid coatings with higher crosslinking density, enhancing abrasion resistance and mechanical strength.

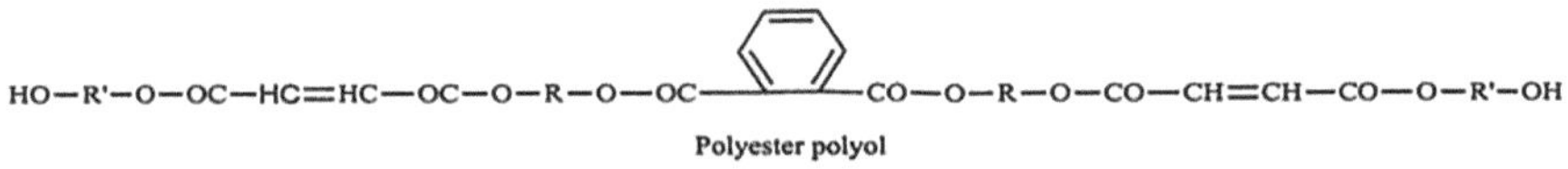

Fig. 2.1 Chemical structure of polyester polyol [2]

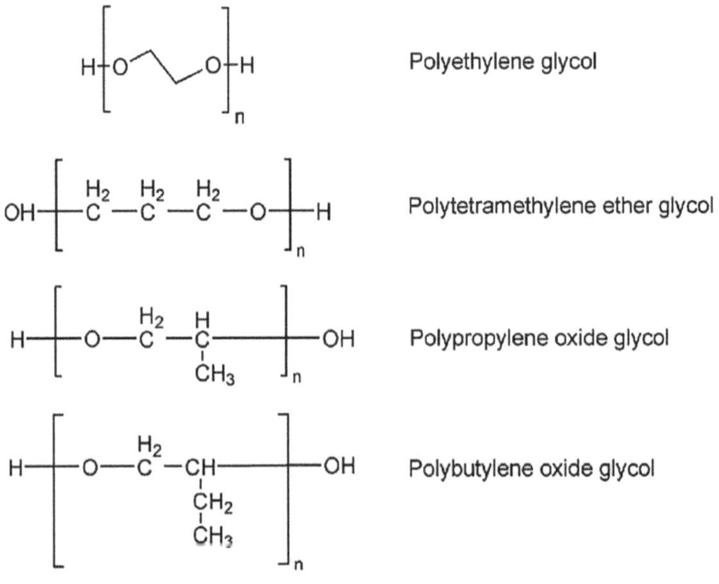

Fig. 2.2 Examples of polyether polyols [4]

This tunability allows polyurethane coatings to be tailored for specific performance requirements across a wide range of industries [3].

2.1.2   Isocyanates

Isocyanates are the reactive components in polyurethane coatings that contain isocyanate (-N=C=O) groups, which react with the hydroxyl groups of polyols to form urethane bonds during the polymerization process. They are broadly classified into two categories:

- **Aromatic isocyanates** (e.g., toluene diisocyanate (TDI), methylene diphenyl diisocyanate (MDI)) are more economical and provide good mechanical strength, but have poor UV stability, making them suitable primarily for indoor applications.
- **Aliphatic isocyanates** (e.g., hexamethylene diisocyanate (HDI), isophorone diisocyanate (IPDI)) offer excellent UV and weathering resistance, ideal for outdoor applications where color and gloss retention are critical.

To achieve the desired crosslinking density, mechanical strength and durability in PU coatings, the isocyanate-to-polyol ratio (NCO/OH) must be carefully controlled, as deviations can lead to reduced performance or incomplete curing [5]. Some common isocyanate structures used in PU manufacturing are shown in Fig. 2.3.

2.1.3   Catalysts

Catalysts are essential components in polyurethane (PU) coatings that accelerate the polymerization reaction between polyols and isocyanates, ensuring efficient curing and uniform film formation. They play a critical role in controlling reaction rate, curing time, film hardness, and final performance of the coating.

There are two primary types of catalysts used in PU coatings:

- **Organometallic catalysts** (e.g., dibutyltin dilaurate, DBTDL) promote the formation of a robust, highly crosslinked polymer network, enhancing mechanical strength, adhesion, and durability.
- **Tertiary amine catalysts** (e.g., triethylamine) accelerate the initial curing and are often used when fast drying or rapid film formation is needed.

In general, amine catalysts are preferred for fast curing applications, while organometallic catalysts are used where long-term durability and structural integrity are priorities. Careful catalyst selection and dosage control are crucial to avoid issues such as brittleness (from excessive crosslinking) or soft films (from undercuring) [6]. Some common catalysts used in PU manufacturing are shown in Fig. 2.4.

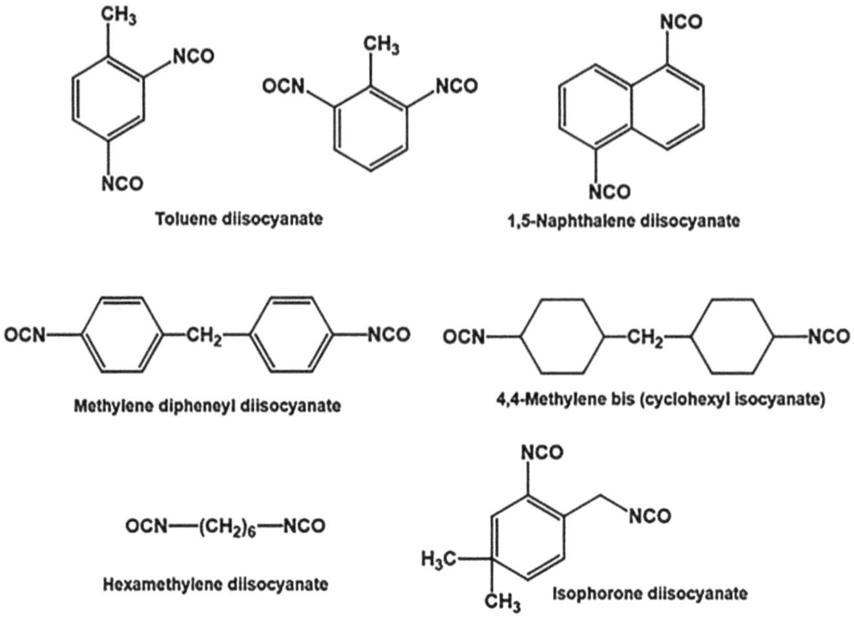

Fig. 2.3 Common isocyanates for polyurethane manufacturing [6]

2.1.4 Solvents

Solvents are key components in polyurethane (PU) coatings, primarily used to adjust the viscosity of the liquid coating, ensuring compatibility with different application methods such as brushing, rolling, or spraying. They also play a critical role in controlling the drying time, which depends on the volatility or vapor pressure of the selected solvent. Common solvents in PU coatings include butyl acetate, xylene, and toluene, which facilitate uniform film formation and efficient drying on the substrate. However, the solvents in solvent-based PU coatings contribute to volatile organic compound (VOC) emissions, which are environmentally harmful and potentially carcinogenic. In contrast, water-based PU coatings use water or eco-friendly solvents, significantly reducing environmental impact while maintaining comparable performance and film quality [7, 8].

2.2 Common Additives in PU Coatings

To tailor polyurethane (PU) coatings for specific application requirements, various additives are incorporated into the formulation. For instance:

Anti-shrinkage Agent: In two-component polyurethane coatings, the molecular weight of the two components is often not high, and they are typically low

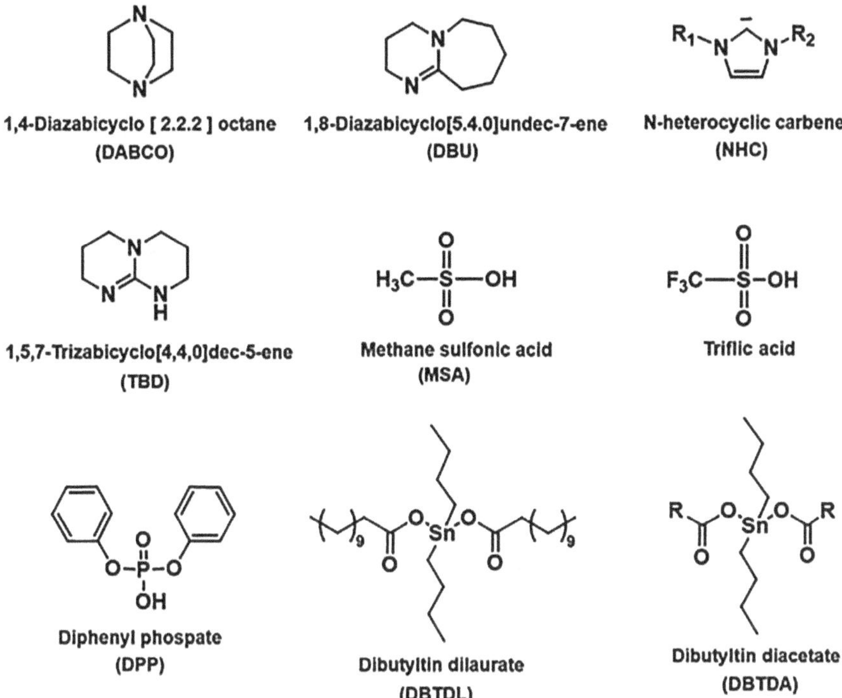

Fig. 2.4 Some common catalysts for polyurethane [6]

oligomers. These components exhibit high polarity and sensitivity to trace oil contamination. Due to differences in surface tension between the two components, the low molecular weight leads to slow film setting (gelation) after application, resulting in a still-flowing liquid state. At this stage, driven by the gradient in surface tension, pinholes are easily formed, necessitating the addition of an anti-shrinkage agent.

Defoaming Agent for Polyurethane Paints: Polyurethane paints often suffer from foaming issues, requiring the addition of defoaming agents. These defoaming agents are typically categorized into non-silicone resin-based and organosilicon-based additives. Resin-based additives consist of thermoplastic copolymers, such as copolymers of ethylene vinyl ether and acrylic esters. These additives, being incompatible with polyurethane paint, disrupt the surface layer of small air bubbles in the coating, causing them to coalesce into larger bubbles over time.

Light Stabilizers: Aliphatic polyurethane clear coatings find extensive use in vehicles and other applications. However, they are prone to cracking and peeling under exposure to sunlight's ultraviolet (UV) radiation. To mitigate this, it is necessary to incorporate light stabilizers, including two main components: ultraviolet absorbers and hindered amine light stabilizers (HALS). The combination of these two components can significantly extend the lifespan of the clear coating.

Moisture scavenger: In polyurethane coatings, the reaction of polyisocyanates with water, especially in the presence of atmospheric moisture, can generate urea and carbon dioxide. This reaction leads to swelling of the coating container, foaming of the film, and an increase in viscosity, ultimately resulting in gelling. Therefore, it is crucial to control the moisture content in the coating. This is particularly important when manufacturing moisture-cured coatings and solvent-free coatings, where measures must be taken to eliminate the moisture introduced by pigments, solvents, and other components. The strategic selection and incorporation of additives allow PU coatings to meet functional, mechanical, and regulatory demands across diverse industries [9].

2.2.1 Chain Extenders and CrossLinking Agents

Additional isocyanates or polyfunctional amines are used as crosslinking agents to enhance the network density of polymer to achieve high chemical and mechanical resistance [1]. Chain extenders, such as 1,4-butanediol or ethylene glycol, are used to increase the molecular weight of polyurethane chains, enhancing mechanical properties like tensile strength and hardness [10]. Crosslinkers, such as trimethylolpropane (TMP) or polyfunctional amines, introduce additional branching, improving coating durability and chemical resistance [11]. Recent studies have explored bio-based chain extenders, such as glycerol derived from biodiesel production, to improve sustainability without compromising performance [12]. The choice of chain extender or crosslinker significantly influences the coating's final properties, requiring careful optimization during formulation.

2.2.2 UV Stabilizers

UV stabilizers added in the recipe to protect the coatings from photodegradation upon exposure to the UV light. These are mostly used in outdoor applications of PU coatings.

2.2.3 Fillers

Fillers are mostly used in the recipe to enhance mechanical properties such as strength or to impart some specific functionality such as opacity by adding titanium dioxide. Most of the time fillers are also added in the coating to reduce the cost of the coating by adding economical fillers such as calcium carbonate [13].

2.2.4 *Thickeners and Rheology Modifiers*

Thickeners used in polyurethane (PU) coatings play a crucial role in enhancing the viscosity and stability of the formulations. Various types of thickeners have been developed, each with unique properties and applications. One of the types of thickeners used in PU coatings is associative thickeners. These thickeners create a thixotropic structure, improving viscosity at low shear rates while maintaining low viscosity during application and often consist of polyethylene glycol, diisocyanate, and end-capping reagents, which allow them to participate in UV curing processes, enhancing stability and performance [14]. Other type is PU-based thickeners, these are synthesized through reactions involving polyether polyols and diisocyanates, resulting in a prepolymer that can effectively increase viscosity in emulsion paints [15].

2.2.5 *Antioxidants*

Polyurethane coatings are susceptible to degradation when exposed to oxygen, elevated temperatures, and ultraviolet radiation, which can significantly reduce their mechanical performance and long-term durability. This degradation primarily occurs through oxidative reactions that generate hydroperoxides and free radicals, leading to chain scission and loss of structural integrity. The incorporation of antioxidants into the coating formulation mitigates these effects by decomposing hydroperoxides and scavenging free radicals, thereby interrupting the oxidation cycle. Consequently, the use of antioxidants not only preserves the mechanical properties of polyurethane coatings but also enhances their service life and storage stability [16].

2.2.6 *Antimicrobial Agents*

Silver, being a noble metal, has gained significant attention in biomedical applications owing to its outstanding antibacterial activity. Silver nanoparticles (AgNPs) and their released Ag^+ ions exert bactericidal effects through multiple mechanisms, including disruption of bacterial cell membranes, penetration into microbial cells, and subsequent interference with intracellular processes such as DNA denaturation and enzymatic dysfunction. This multifaceted mode of action results in effective inhibition of bacterial growth and survival. Furthermore, the incorporation of alternative nanoparticles, such as other noble metals (e.g., Au, Cu), metal oxides (e.g., TiO_2, ZnO), and metal phosphides into polyurethane (PU) nanocomposites has also been demonstrated to significantly enhance their antibacterial performance, thereby broadening the scope of PU-based functional coatings for biomedical and hygienic applications [17].

2.2.7   Anti-foaming Agents

Defoaming agents, also referred to as anti-foaming agents, are essential additives in polyurethane coating formulations, designed to eliminate or suppress foam formation during processing and application. Foam can negatively affect film uniformity, adhesion, and surface appearance, leading to defects in the final coating. These agents function by reducing the surface tension gradient within the liquid coating, thereby collapsing bubbles and preventing their stabilization. Generally, defoaming agents are classified into two categories: non-silicone resin-based and organosilicon-based additives. While non-silicone types are effective for minimizing surface defects without influencing intercoat adhesion, organosilicon defoamers are widely utilized for their strong and long-lasting defoaming efficiency. The appropriate selection and dosage of defoaming agents are crucial to achieving smooth, defect-free coatings with improved surface texture and overall performance [7].

2.2.8   Flame Retardant

A flame-retardant system is a compound or formulation incorporated into polymer substrates to enhance their resistance to ignition and combustion. Effective flame-retardant function by interfering with one or more of the elements are required to sustain flammability heat, fuel, or oxygen. Their primary purpose is to reduce the inherent fire risk of polymers by slowing down flame propagation under fire conditions. Mechanistically, flame retardants disrupt the thermal decomposition pathway of polymers, and their effectiveness often depends on both their chemical nature and compatibility with the specific substrate. The use of halogen-containing flame retardants became widespread in the 1970s, with brominated systems gaining particular prominence in the 1980s. These retardants act primarily in the gas phase: upon heating, they decompose into halogen radicals that inhibit free-radical chain reactions, thereby obstructing the oxidation of volatile fuels. This suppression lowers the concentration of oxygen available for combustion, effectively extinguishing the flame. Their efficiency can be further enhanced by synergistic compounds such as antimony- and phosphorus-based additives, which facilitate free-radical scavenging and regeneration of halogen radicals.

However, by the 1990s, concerns over environmental and health hazards associated with halogen-based systems including the release of toxic and corrosive gases (e.g., hydrogen chloride and hydrogen bromide), excessive smoke production, and disposal challenges prompted a shift toward halogen-free flame retardants. Current research increasingly emphasizes environmentally benign systems, though these alternatives are often less efficient and more costly compared to their halogenated counterparts. Several non-halogen flame-retardant mechanisms have been developed. Some formulations release non-combustible gases (e.g., water vapor, CO_2, or ammonia), which dilute the concentration of fuel or oxygen, reducing flame intensity. Others promote the formation of protective char layers or solid residues on the polymer surface, thereby slowing heat transfer and reducing the overall heat release rate. A notable class within this category is intumescent flame retardants, which expand to form an insulating foam char when exposed to heat. This barrier limits oxygen diffusion, reduces flammable volatiles, and protects the underlying polymer. In many cases, flame retardants may operate through a combination of these mechanisms to achieve optimal fire resistance [18].

2.3 Processing and Application Considerations

Formulation guidelines for polyurethane coatings extend beyond the selection of raw materials to encompass processing parameters and application techniques. Factors such as the mixing sequence, application temperature, and ambient humidity have a profound impact on the final coating performance. In two-component (2K) systems, strict control of the stoichiometric ratio between polyol and isocyanate is essential. Any deviation may result in incomplete curing, residual unreacted species, or an excess of isocyanate, which can compromise the coating by increasing brittleness and reducing long-term durability [19]. Temperature and humidity play a crucial role in the curing kinetics and final properties of polyurethane coatings. Under high humidity conditions (e.g., $\geq 75\%$ RH), isocyanate groups can react with ambient moisture, leading to excessive urea formation. This often results in defects such as foaming, surface roughness, and reduced gloss. In contrast, low humidity environments generally favor the formation of smoother films but may cause stratification within the coating layer. This stratification is typically observed as differences in the glass transition temperature (Tg) between the film–air interface and the film–substrate interface, ultimately affecting mechanical uniformity and performance [19]. Application techniques such as spraying, brushing, or roll-coating require careful adjustment of the coating's viscosity to ensure uniform film formation and optimal performance. For instance, spray applications demand relatively low viscosities, which are commonly achieved by solvent dilution in solvent-borne systems or through formulation adjustments in waterborne systems. In contrast, brushing and roll-coating can tolerate higher viscosities, allowing for thicker film build in a single application. Therefore, tailoring viscosity to the selected application method is critical for achieving consistent coating quality and minimizing defects such as sagging, orange peel, or poor leveling [20]. Curing conditions, particularly temperature and time, must be carefully optimized to ensure complete crosslinking and defect-free film formation. Improper curing can lead to issues such as bubbling, surface irregularities, or cracking, which compromise coating performance. Traditionally, thermal curing has been employed; however, recent advancements have introduced UV-curable polyurethane (PU) coatings, which enable rapid curing at ambient temperature, significantly reducing energy consumption and production cycle times. Despite these advantages, UV-curable systems require specialized photoinitiators and equipment, and their performance can be limited by the penetration depth of UV light in highly pigmented or opaque coatings [21].

2.4 Emerging Trends and Innovations

Recent advancements in polyurethane coatings focus on the development of smart and functional properties that extend beyond traditional protective roles. Among these, self-healing PU coatings have gained considerable attention. By incorporating microcapsules containing healing agents or designing reversible covalent/noncovalent bonds within the polymer matrix, these coatings can autonomously repair scratches or microcracks when damage occurs. This intrinsic self-repair capability not only restores barrier performance but also significantly prolongs the service life of the coated substrate, thereby reducing maintenance costs and enhancing sustainability [22]. Stimuli-responsive polyurethane coatings represent a significant advancement, as they can adapt their structure or functionality in response to external stimuli such as temperature, pH, light, or electrical fields. These coatings are being designed for nanotechnology has greatly enhanced the performance of polyurethane (PU) coatings. PU nanocomposites exhibit superior thermal, mechanical, and physicochemical properties compared to neat polyurethane due to the incorporation of nanoparticles. A major advancement in this area is the development of nanocomposite coatings reinforced with nanocarbon particles such as carbon nanotubes, graphene, graphene oxide (GO), graphene nanoribbons, and nanodiamonds as well as inorganic nanoparticles. These nanofillers significantly improve weathering resistance, solvent and chemical resistance, pH stability, wear resistance, and overall mechanical strength. Owing to these enhanced characteristics, PU nanocomposite coatings are increasingly applied in corrosion protection, textiles, biomedical devices, and defense technologies including biomedical devices, smart packaging, and protective systems, where dynamic adaptability offers enhanced performance compared to conventional coatings [23]. Nanotechnology has greatly enhanced the performance of polyurethane (PU) coatings. PU nanocomposites exhibit superior thermal, mechanical, and physicochemical properties compared to neat polyurethane due to the incorporation of nanoparticles. A major advancement in this area is the development of nanocomposite coatings reinforced with nanocarbon particles such as carbon nanotubes, graphene, graphene oxide (GO), graphene nanoribbons, and nanodiamonds as well as inorganic nanoparticles. These nanofillers significantly improve weathering resistance, solvent and chemical resistance, pH stability, wear resistance, and overall mechanical strength. Owing to these enhanced characteristics, PU nanocomposite coatings are increasingly applied in corrosion protection, textiles, biomedical devices, and defense technologies [24].

2.5 Application Techniques

Polyurethane (PU) coatings are highly regarded for their durability, versatility, and superior mechanical performance, making them indispensable across industries such as automotive, construction, and aerospace. Their final properties are governed not only by formulation parameters but also by the application techniques employed, including conventional methods like spraying, brushing, dipping, and roll-coating, as well as advanced approaches such as electrostatic spraying. Some PU coating techniques for different applications are mentioned in Fig. 2.5. Each technique directly impacts film thickness, surface uniformity, adhesion, and overall coating quality. This literature review explores the major application methods for PU coatings, analyzing their advantages, limitations, and recent advancements, with particular emphasis on optimizing performance while addressing environmental and sustainability challenges.

2.5.1 Spraying Techniques

Spraying is among the most widely adopted methods for applying polyurethane coatings, valued for its efficiency and ability to uniformly coat complex geometries. In conventional air spraying, compressed air atomizes the coating into fine droplets, producing a smooth and uniform film. This technique is particularly effective for large surface areas, such as automotive bodies; however, it often results in significant overspray, contributing to material loss and elevated volatile organic compound (VOC) emissions. To address these limitations, high-volume low-pressure (HVLP) spraying has been developed, which employs reduced air pressure to enhance

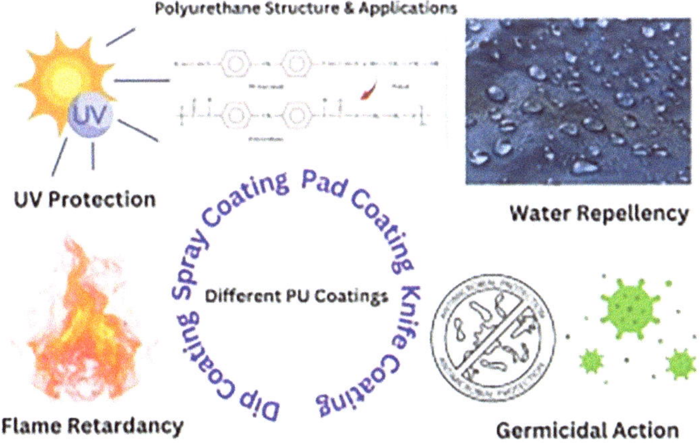

Fig. 2.5 Different application techniques of PU coatings for different applications [25]

transfer efficiency, thereby minimizing overspray and lowering environmental impact [26].

Airless spraying, which utilizes high-pressure pumps to atomize the coating without compressed air, is particularly suitable for high-viscosity polyurethane formulations, such as two-component (2K) systems. This method enables the application of thicker films in a single pass, improving productivity and ensuring strong protective performance [27]. However, precise control of pressure and nozzle size in airless spraying is critical to prevent surface defects such as orange peel or sagging. Recent advancements include electrostatic spraying, in which charged coating particles are directed toward a grounded substrate, significantly enhancing transfer efficiency and reducing overspray, particularly on metal substrates. This technique is extensively applied in the automotive and appliance industries, though its use is limited by the requirement for conductive substrates and the need for specialized equipment [28].

2.5.2 Brushing Techniques

Brushing remains a traditional and cost-effective technique for applying polyurethane coatings, particularly in small-scale or touch-up applications such as furniture finishing and architectural surfaces. This method offers precise control over film thickness and is compatible with both solvent-based and waterborne formulations. However, brushing is often limited by the potential for surface defects, including visible brush marks and film non-uniformity, especially when working with high-viscosity polyurethane systems [29]. The selection of brush material plays a critical role in coating quality. Natural bristle brushes are generally preferred for solvent-based polyurethane coatings due to their ability to hold and release viscous formulations smoothly, while synthetic bristles are more suitable for waterborne systems, minimizing swelling and maintaining consistent film formation [30].

Formulation adjustments, including the incorporation of leveling agents and rheology modifiers, are essential to minimize brush marks and promote uniform, smooth film formation [31]. Recent studies have demonstrated that incorporating associative thickeners into water-based polyurethane coatings can significantly improve flow and leveling during brushing, thereby enhancing both aesthetic appearance and functional performance. However, brushing is less efficient for large surface areas and often requires multiple coats to achieve the desired film thickness, which increases application time and labor costs [32].

2.5.3 Dipping Techniques

Dipping is a straightforward application technique in which the substrate is immersed in a polyurethane coating bath, forming a uniform film upon withdrawal. This method is particularly effective for coating complex geometries, such as wires, pipes, and small components, and is widely employed in industries including electronics and medical devices. The resulting film thickness is primarily governed by the withdrawal speed and the viscosity of the coating; slower withdrawal rates generally yield thinner films [33].

Dipping is particularly suitable for one-component (1K) polyurethane systems, such as moisture-cured coatings, owing to their straightforward application process and extended pot life [34]. In contrast, two-component systems require stringent control of pot life to avoid premature curing within the bath. Key challenges include managing dripping and achieving uniform film thickness on vertical surfaces, which can be addressed by optimizing formulation viscosity and incorporating defoaming agents [35].

2.5.4 Roll-Coating Techniques

This technique operates by transferring adhesive from a trough via a pickup roller, which is partially immersed in the adhesive, onto a contacting transfer roller sheet. The substrate is continuously coated as it passes between the transfer roller and a pressure roller, with the latter adjusted to control the desired coating thickness. Roll-coating is particularly suitable for applying adhesives to flat sheets and films and can accommodate parts up to 1.83 m in size. Under optimal conditions, it delivers high production rates and uniform coverage. For smaller components, it is often more economical to roll-coat large sheets and subsequently stamp out the required parts, as the adhesive waste generated is less costly than coating small individual items [36]. Reverse roll-coating, in which the coating is applied counter to the direction of substrate movement, is highly effective for producing high-gloss finishes, particularly in furniture and flooring applications [37].

Roll-coating is compatible with both solvent-based and water-based polyurethane systems; however, water-based formulations often require extended drying times or the use of heated rollers to minimize defects such as blistering. The process provides high transfer efficiency and reduced material waste compared to spraying, but its applicability is restricted to flat or moderately curved surfaces [26]. Recent advancements involve the utilization of microstructured rollers to produce textured or patterned coatings, thereby broadening the aesthetic and functional possibilities for decorative applications [38].

2.5.5 Curtain Coating

Curtain coating is a high-throughput technique in which a substrate passes through a continuous curtain of liquid coating, enabling uniform coverage with precise control over film thickness. It is particularly suited for flat substrates such as wood panels and metal sheets, making it widely applicable in furniture and industrial manufacturing. Successful operation requires stringent control of flow rate and viscosity to avoid defects such as edge buildup and curtain instability. Although less adaptable to complex geometries compared to spraying or dipping, curtain coating remains highly efficient for large-scale production [39]. This technique is compatible with both high solid and water-based polyurethane formulations, thereby contributing to reduced VOC emissions [35].

2.5.6 Emerging Application Techniques

Recent advancements in application technology have broadened the capabilities of polyurethane coatings. UV-curable polyurethane systems, applied through spraying or roll-coating, enable rapid curing and reduced energy consumption, rendering them highly suitable for high-speed production lines. The incorporation of photoinitiators facilitates curing within seconds under UV irradiation; however, this method necessitates specialized equipment and is generally restricted to clear or lightly pigmented formulations [40].

Powder coating with polyurethane resins is an emerging technique that provides zero-VOC emissions and exceptional durability, particularly for metal substrates. The process involves the electrostatic deposition of powdered polyurethane, followed by thermal curing, and has gained increasing adoption in the automotive and appliance industries. However, challenges such as maintaining uniform particle size and managing the high curing temperatures required can limit compatibility with certain substrates [41].

2.5.7 Environmental and Processing Considerations

The selection of an application technique is strongly influenced by environmental regulations, particularly those concerning VOC emissions. Spraying, although versatile, can contribute substantially to VOC release unless low-VOC or water-based formulations are employed. In contrast, techniques such as roll-coating and curtain coating provide higher transfer efficiency, thereby minimizing material waste and emissions. Water-based systems, however, demand strict control of drying conditions to prevent defects such as cracking or inadequate adhesion, often necessitating the use of heated drying systems. Regardless of the technique, substrate preparation is critical to ensuring coating adhesion and long-term performance, with processes such as surface cleaning, sanding, or priming being especially important for metal and plastic substrates [42].

Control of humidity and temperature during application and curing is essential, particularly for moisture-cured and two-component polyurethane systems, to prevent defects such as bubbling, incomplete curing, or compromised film integrity [43].

2.5.8   Surface Preparation and Priming Before Coating Application

Surface preparation and priming are vital steps in the application of polyurethane coatings, as they directly affect adhesion, durability, and overall performance. Effective surface preparation ensures that the substrate is clean, smooth, and chemically compatible with the coating, while the application of a primer enhances interfacial bonding and provides additional resistance against environmental degradation [44]. These processes are critical across diverse industries including automotive, aerospace, construction, and marine, where polyurethane coatings are highly valued for their versatility, durability, and superior mechanical performance [45]. The following sections discuss the techniques, materials, and recent advancements in surface preparation and priming for polyurethane coatings, with emphasis on their influence on coating performance and sustainability.

2.6   Importance of Surface Preparation

Surface preparation forms the foundation of successful polyurethane coating applications, as it eliminates contaminants, enhances surface energy, and improves adhesion. Impurities such as oils, grease, dust, rust, or residual coatings can significantly compromise interfacial bonding, often resulting in defects such as delamination or blistering. The primary objective is to achieve a clean, properly roughened surface that maximizes both mechanical interlocking and chemical bonding between the substrate and the coating [46].

The selection of an appropriate surface preparation technique is determined by factors such as substrate material (e.g., metal, plastic, wood, or concrete), coating formulation, and application environment. Typical substrates for polyurethane coatings include steel, aluminum, plastics, and wood, each of which requires tailored preparation strategies to ensure optimal adhesion. Insufficient or improper preparation can result in premature coating failure, leading to increased maintenance costs and reduced service life.

2.6.1   Mechanical Methods

Mechanical methods such as abrasive blasting, sanding, and grinding are commonly employed to clean and roughen substrates prior to polyurethane coating application. Abrasive blasting techniques, including sandblasting and shot blasting, effectively remove rust, scale, and existing coatings from metal surfaces while generating a textured profile that enhances mechanical adhesion. The surface roughness (Ra) plays a critical role in coating performance; increased roughness generally improves adhesion for thick polyurethane layers used in industrial applications. However, excessively rough surfaces may entrap air or contaminants, potentially leading to coating defects [47].

Sanding is widely employed for wood and plastic substrates, as it removes surface imperfections, increases surface area, and enhances bonding with polyurethane coatings [48]. For plastic substrates, which typically exhibit low surface energy, sanding or abrasion is often complemented by chemical surface treatments to enhance wettability and promote stronger adhesion of polyurethane coatings [49]. Grinding is commonly used for concrete substrates to remove laitance and open surface pores, thereby facilitating strong adhesion of polyurethane floor coatings [50].

2.6.2   Chemical Surface Cleaning

Chemical cleaning utilizes solvents, detergents, or alkaline and acidic solutions to eliminate organic contaminants such as oils and grease. In the case of metal substrates, alkaline cleaners are particularly effective for degreasing, whereas acid pickling is commonly employed to remove oxides, scale, and other inorganic residues, thereby improving coating adhesion [51]. Solvent cleaning, typically using acetone or isopropanol, is widely applied for plastics and composites; however, careful solvent selection is essential to prevent substrate damage or degradation [52]. Environmental considerations have led to the increased adoption of water-based cleaners and biodegradable solvents, aimed at minimizing volatile organic compound (VOC) emissions [53]. For polyurethane coatings, chemical cleaning must be compatible with the coating chemistry. For instance, residual alkaline cleaners can interfere with the curing of two-component (2K) polyurethane systems, making thorough rinsing essential to prevent coating defects [11].

2.6.3   Surface Activation

Low surface energy substrates, such as plastics (e.g., polyethylene and polypropylene), require surface activation to enhance wettability and adhesion. Common activation techniques include plasma treatment, corona discharge, and flame treatment, all of which introduce polar functional groups (e.g., hydroxyl, carboxyl) onto the surface. Plasma treatment, utilizing ionized gases such as oxygen or nitrogen, is particularly effective for polymers, as it increases surface energy without affecting the bulk properties of the material [54]. Corona discharge is extensively employed in industrial settings for the continuous treatment of plastic films, whereas flame treatment is typically preferred for three-dimensional components, providing rapid surface activation and improved adhesion. These surface activation techniques are especially critical for water-based polyurethane dispersions (PUDs), which demand high substrate surface energy to ensure proper wetting and uniform film formation. Recent advancements include atmospheric plasma systems, providing cost-effective and scalable solutions for industrial surface activation [55].

2.6.4   Degreasing and Decontamination

Degreasing is a crucial step for both metal and plastic substrates to eliminate oils, fingerprints, and other surface contaminants. Historically, vapor degreasing with chlorinated solvents was widely used, but environmental regulations have largely shifted the industry toward aqueous or semi-aqueous cleaning systems. Ultrasonic cleaning, employing high-frequency sound waves in a liquid medium, is increasingly applied for precision components, such as aerospace parts, to remove microscopic contaminants. Complete removal of cleaning agents is essential to avoid interference with the curing of polyurethane coatings [56, 57].

2.6.5   Priming for Polyurethane Coatings

Primers are intermediate layers applied following surface preparation to improve adhesion, corrosion resistance, and overall substrate compatibility. In polyurethane coating systems, primers are specifically formulated to match both the substrate and the topcoat chemistry, ensuring the formation of a robust interfacial bond [58]. Primers also impart additional functionalities, including corrosion protection for metal substrates [109] and sealing for porous materials such as wood or concrete.

2.6.5.1   Epoxy Primers

Epoxy primers are commonly employed for metal substrates, such as steel and aluminum, owing to their superior adhesion and corrosion resistance. They provide a robust barrier against moisture and chemical exposure, making them particularly suitable for marine and industrial applications. Epoxy primers are frequently paired with polyurethane topcoats to form durable, multi-layer coating systems. However, their relatively slower curing times compared to polyurethane primers can constrain production efficiency [59].

2.6.5.2   Polyurethane Primers

Polyurethane primers, available as one-component (1K) or two-component (2K) systems, are employed when compatibility with the polyurethane topcoat is essential. They offer excellent adhesion and flexibility, making them particularly suitable for plastics and composite substrates [32]. Moisture-cured polyurethane primers are particularly effective for concrete substrates, as they seal pores and prevent outgassing during the application of the topcoat [60]. A polyurethane primer typically consists of a base material and a curing agent. The base material comprises 40–60 wt% of a polymer–polyol mixture, 15–25 wt% solvent, 10–20 wt% filler, and 5–15 wt% additives, relative to the total base component. The filler may include one or more materials selected from alumina powder, stone powder, and hygroscopic agents. Additives consist of chloroprene rubber combined with solvent in a 1:1 weight ratio. The curing agent is composed of polyethylene polyphenyl-isocyanate. The polymer–polyol mixture itself contains a combination of base polyol and polymer polyol in an 85:15 wt% ratio [61]. Recent developments in polyurethane primers include low-VOC formulations, designed to comply with environmental regulations and reduce volatile organic compound emissions [62].

2.6.5.3   Silane and Zinc-Rich Primers

Silane-based primers are applied to glass and metal substrates, where they form covalent bonds that significantly enhance adhesion [63]. Zinc-rich primers, formulated with high concentrations of zinc dust, offer cathodic protection for steel substrates and are commonly employed in harsh environments such as marine or offshore structures [64]. Such primers require careful formulation to maintain compatibility with polyurethane topcoats, as the high zinc content can adversely affect curing and overall coating performance [65].

2.7   Primer Application Techniques

Primers are applied using various techniques, including spraying, brushing, or dipping, depending on the substrate type and production requirements. Spraying is generally preferred for large surfaces, providing uniform coverage, whereas brushing is typically employed for touch-ups or smaller areas [34]. Dipping is an effective application method for small components, such as fasteners, but requires precise control of primer viscosity to prevent excessive coating buildup [31]. Electrostatic spraying is increasingly employed for primer application, enhancing transfer efficiency and minimizing material waste [66].

2.8   Recent Advancements and Future Directions

Recent advancements in surface preparation and priming focus on enhancing efficiency, sustainability, and coating performance. Automated systems, including robotic abrasive blasting and plasma treatment, improve process consistency while reducing labor requirements and associated costs [67]. Nanotechnology is increasingly impacting primer development, with nanoparticle-enhanced formulations offering improved corrosion resistance and adhesion. For instance, the incorporation of silica or graphene oxide nanoparticles into epoxy primers has demonstrated superior barrier and protective properties [68].

Smart primers with self-healing or corrosion-sensing functionalities are emerging, utilizing microcapsules or indicators that respond dynamically to environmental changes [69]. Machine learning techniques are increasingly applied to optimize surface preparation parameters, enabling prediction of adhesion performance based on substrate characteristics and coating properties [70]. Collectively, these innovations are expected to significantly enhance the reliability and service life of polyurethane coating systems.

2.9 Challenges and Considerations

Despite recent advancements, challenges persist in surface preparation and priming. Inconsistent surface cleaning can result in adhesion failures, particularly on complex substrates such as composites [71]. Excessive priming can lead to overly thick films, causing cracking or reduced flexibility, whereas insufficient priming may result in inadequate adhesion and compromised coating performance [72]. Balancing cost, performance, and environmental impact remain a significant challenge, as eco-friendly methods such as water-based cleaning or laser ablation often entail higher operational costs [73].

References

1. Chattopadhyay DK, Raju KVSN (2007) Structural engineering of polyurethane coatings for high performance applications. Prog Polym Sci 32(3):352–418. https://doi.org/10.1016/j. progpolymsci.2006.05.003
2. Synthesis of diphosphorus-based polyurethane esters and their application in flame-retardant nanoclay coatings. https://www.researchgate.net/publication/326355848_Synthesis_of_ diphosphorus-based_polyurethane_esters_and_their_application_in_flame-retardant_nano- clay_coatings. Accessed 21 Aug 2025
3. Hepburn C (2012) Polyurethane elastomers. Springer Science & Business Media
4. Arévalo-Alquichire S, Valero M, Arévalo-Alquichire S, Valero M (2017) Castor oil polyure- thanes as biomaterials. In: Elastomers. IntechOpen. https://doi.org/10.5772/intechopen.68597
5. Polyurethane formulation: the chemistry of isocyanate and polyols. https://doxuchem.com/ polyurethane-formulation-the-chemistry-of-isocyanate-and-polyols/. Accessed 29 July 2025
6. de Souza FM, Kahol PK, Gupta RK (2021) Introduction to polyurethane chemistry. In: Polyurethane chemistry: renewable polyols and isocyanates, ACS symposium series, vol 1380. American Chemical Society, pp 1–24. https://doi.org/10.1021/bk-2021-1380.ch001
7. Solvents and additives of polyurethane coatings. Doxu Chemical. https://doxuchem.com/ solvents-and-additives-of-polyurethane-coatings/. Accessed 29 July 2025

8. Recent developments in aqueous two-component polyurethane (2K-PUR) coatings. https://www.researchgate.net/publication/222119884_Recent_developments_in_aqueous_two-component_polyurethane_2K-PUR_coatings. Accessed 29 July 2025

9. Solvents and additives of polyurethane coatings. Doxu Chemical. https://doxuchem.com/solvents-and-additives-of-polyurethane-coatings/. Accessed 15 Aug 2025

10. Ionescu M (2005) Chemistry and technology of polyols for polyurethanes. Rapra Technology. https://www.google.com/search?q=%E2%80%A2+Ionescu%2C+M.+(2005).+Chemistry+and+Technology+of+Polyols+for+Polyurethanes.+Rapra+Technology.&rlz=1C1FHFK_en-GBPK1170PK1170&oq=%E2%80%A2%09Ionescu%2C+M.+(2005).+Chemistry+and+Technology+of+Polyols+for+Polyurethanes.+Rapra+Technology.&gs_lcrp=EgZjaHJvbWUyCAgAEEUYJxg50gEHNzMxajBqN6gCALACAA&sourceid=chrome&ie=UTF-8. Accessed 4 Aug 2025

11. Meier-Westhues U (2007) Polyurethanes: coatings, adhesives and sealants. Vincentz Network

12. Paraskar PM, Kulkarni RD (2022) Influence of bio-based chain extender glycerol on the performance of dimer fatty acid-derived polyurethane coatings. J Polym Res 29:12. https://doi.org/10.1007/s10965-022-02964-0

13. Deshmukh GP, Mahanwar PA (2019) Review on the use of nanofillers in polyurethane coating systems for different coating applications. J Coat Sci Technol 6(1):22–35. https://doi.org/10.6000/2369-3355.2019.06.01.3

14. Dan S (2017) Associated thickener for waterborne UV polyurethane coating and preparation method thereof. https://scispace.com/papers/associated-thickener-for-waterborne-uv-polyurethane-coating-4zmif6p7x5. Accessed 5 Aug 2025

15. Nan G et al (2015) Preparation of associative polyurethane thickener and its thickening mechanism research. J Nanomater. https://doi.org/10.1155/2015/137646

16. Polyurethane applications: the role of antioxidants in ensuring stability. Chic. https://chicchem.com/polyurethane-applications-the-role-of-antioxidants-in-ensuring-stability/. Accessed 29 July 2025

17. Kasi G, Gnanasekar S, Zhang K, Kang ET, Xu LQ (2022) Polyurethane-based composites with promising antibacterial properties. J Appl Polymer Sci 139(20):52181. https://doi.org/10.1002/app.52181

18. Mohd Sabee MMS et al (2022) Flame retardant coatings: additives, binders, and fillers. Polymers 14(14):14. https://doi.org/10.3390/polym14142911

19. Liu Y et al (2024) Stratification of polyisocyanate in two-component waterborne polyurethane films. Chem Eng J 483:148981. https://doi.org/10.1016/j.cej.2024.148981

20. Talbert R. Paint technology handbook. https://www.perlego.com/book/1628166/paint-technology-handbook-pdf. Accessed 5 Aug 2025

21. Xu H, Qiu F, Wang Y, Wu W, Yang D, Guo Q (2012) UV-curable waterborne polyurethane-acrylate: preparation, characterization and properties. Prog Org Coat 73(1):47–53. https://doi.org/10.1016/j.porgcoat.2011.08.019

22. Wang H, Xu J, Du X, Du Z, Cheng X, Wang H (2021) A self-healing polyurethane-based composite coating with high strength and anti-corrosion properties for metal protection. Compos Part B Eng 225:109273. https://doi.org/10.1016/j.compositesb.2021.109273

23. Fabric coated with shape memory polyurethane and its properties. https://www.mdpi.com/2073-4360/10/6/681. Accessed 6 Aug 2025

24. Kausar A (2018) Polyurethane nanocomposite coatings: state of the art and perspectives. Polym Int 67(11):1470–1477. https://doi.org/10.1002/pi.5616

25. Srivastava A, Maity S, Das BR (2025) A review on multi-functional polyurethane (PU) coatings for fabric applications: materials, processes and recent developments. Prog Org Coat 207:109377. https://doi.org/10.1016/j.porgcoat.2025.109377

26. 4 types of coating techniques. Doxu Chemical. https://doxuchem.com/4-types-of-coating-techniques/. Accessed 5 Aug 2025

27. 2K-PU. https://dispersions-resins.basf.com/emea/en/newsletter-coatings/2018-02/2K-PU. Accessed 5 Aug 2025

28. The propertïes of MWCNT/polyurethane conductive composite coating prepared by electrostatic spraying. https://www.researchgate.net/publication/284141721_The_properties_of_ MWCNTpolyurethane_conductive_composite_coating_prepared_by_electrostatic_spraying. Accessed 5 Aug 2025

29. How to apply industrial coatings: step-by-step application process with cumulus. Cumulus Quality. https://cumulusquality.com/how-to-apply-industrial-coatings-step-by-step-application-process-with-cumulus-pro/. Accessed 5 Aug 2025

30. Bailão A. Brushes for retouching: how to choose them. Conser Vation

31. Waterborne polyurethanes: a review. https://www.researchgate.net/publication/316975981_ Waterborne_Polyurethanes_A_Review. Accessed 5 Aug 2025

32. Waterborne polyurethanes: a review. ResearchGate. https://www.researchgate.net/publication/316975981_Waterborne_Polyurethanes_A_Review. Accessed 5 Aug 2025

33. Brinker CJ (2013) Dip coating. In: Schneller T, Waser R, Kosec M, Payne D (eds) Chemical solution deposition of functional oxide thin films. Springer, Vienna, pp 233–261. https://doi.org/10.1007/978-3-211-99311-8_10

34. Randall D, Lee S (2002) The polyurethanes book. Wiley, New York, pp 1–477. https://www.scirp.org/reference/referencespapers?referenceid=437702. Accessed 6 Aug 2025

35. Noble K-L (1997) Waterborne polyurethanes. Prog Org Coat 32(1–4):131–136. https://doi.org/10.1016/S0300-9440(97)00071-4

36. Roll coating—an overview. ScienceDirect Topics. https://www.sciencedirect.com/topics/engineering/roll-coating. Accessed 6 Aug 2025

37. Roller coating—an overview. ScienceDirect Topics. https://www.sciencedirect.com/topics/engineering/roller-coating. Accessed 6 Aug 2025

38. Jiang L-T, Huang T-C, Chang C-Y, Ciou J-R, Yang S-Y, Huang P-H (2007) Direct fabrication of rigid microstructures on a metallic roller using a dry film resist. J Micromech Microeng 18(1):015004. https://doi.org/10.1088/0960-1317/18/1/015004

39. Curtain coating—an overview. ScienceDirect Topics. https://www.sciencedirect.com/topics/engineering/curtain-coating. Accessed 6 Aug 2025

40. UV curable polyurethane acrylate coatings for metal surfaces. Pigment & Resin Technology, Emerald Publishing. https://www.emerald.com/prt/article-abstract/37/4/217/330344/UV-curable-polyurethane-acrylate-coatings-for?redirectedFrom=fulltext. Accessed 7 Aug 2025

41. Farshchi N, Gedan-Smolka M (2020) Polyurethane powder coatings: a review of composition and characterization. Ind Eng Chem Res 59(34):15121–15132. https://doi.org/10.1021/acs.iecr.0c02320

42. Surface preparation: the essential step before coating. https://vivablast.com/new/what-is-surface-preparation-why-is-it-important-before-coating/. Accessed 7 Aug 2025

43. Competing bubble formation mechanisms in rigid polyurethane foaming. ResearchGate. https://www.researchgate.net/publication/351637469_Competing_bubble_formation_mechanisms_in_rigid_polyurethane_foaming. Accessed 7 Aug 2025

44. Surface preparation: the essential step before coating. https://vivablast.com/new/what-is-surface-preparation-why-is-it-important-before-coating/. Accessed 11 Aug 2025

45. What is surface preparation? (An in-depth guide). https://www.twi-global.com/technical-knowledge/faqs/what-is-surface-preparation.aspx. Accessed 11 Aug 2025

46. Barnhart R, Mericle D, Hocking T. Why surface preparation is important

47. Improving the adhesive properties by sandblasting the surface with copper slag and glass beads. https://www.mdpi.com/1996-1944/18/8/1746. Accessed 11 Aug 2025

48. Wood sanding 101—beginners' guide to sanding wood. Empire Abrasives. https://www.empireabrasives.com/blog/wood-sanding-101-beginners-guide-to-sanding-wood/. Accessed 11 Aug 2025

49. Kruse A, Krüger G, Baalmann A, Hennemann O-D (1995) Surface pretreatment of plastics for adhesive bonding. J Adhes Sci Technol 9(12):1611–1621. https://doi.org/10.1163/156856195X00248

50. Applying polyurethane flooring like a pro. Technical Finishes. https://technicalfinishes.com/applying-polyurethane-flooring-like-a-pro/. Accessed 11 Aug 2025
51. Preparing surfaces for excellence. https://vivablast.com/our-services/protection-solutions/surface-preparation/. Accessed 11 Aug 2025
52. Plastic coating—an overview. ScienceDirect Topics. https://www.sciencedirect.com/topics/materials-science/plastic-coating. Accessed 11 Aug 2025
53. Power of water based cleaners: introduction, advantages, and applications. https://www.zavenir.com/insight-category/general-articles/power-of-water-based-cleaners-introduction-advantages-and-applications. Accessed 11 Aug 2025
54. Liston EM (1989) Plasma treatment for improved bonding: a review. J Adhesion 30(1–4):199–218. https://doi.org/10.1080/00218468908048206
55. Dole N, Ahmadi K, Solanki D, Swaminathan V, Keswani V, Keswani M (2024) Corona treatment of polymer surfaces to enhance adhesion. In: Polymer surface modification to enhance adhesion. Wiley, pp 45–76. https://doi.org/10.1002/9781394231034.ch2
56. A guide to cleaning & degreasing surfaces before painting. https://advancepainting.com.au/cleaning-degreasing-surfaces-guide/. Accessed 7 Aug 2025
57. Degreasing agent—an overview. ScienceDirect Topics. https://www.sciencedirect.com/topics/chemistry/degreasing-agent. Accessed 7 Aug 2025
58. Primer (Paint)—an overview. ScienceDirect Topics. https://www.sciencedirect.com/topics/chemistry/primer-paint. Accessed 5 Aug 2025
59. Liu J et al (2024) Study on the chemical bonding at the interface between epoxy primer and polyurethane topcoat. Prog Org Coat 196:108677. https://doi.org/10.1016/j.porgcoat.2024.108677
60. Progress in waterborne polymer dispersions for coating applications: commercialized systems and new trends. https://www.researchgate.net/publication/385259789_Progress_in_waterborne_polymer_dispersions_for_coating_applications_commercialized_systems_and_new_trends. Accessed 5 Aug 2025
61. Primer for polyurethane. SciSpace. https://scispace.com/papers/primer-for-polyurethane-3ncl2swr0x. Accessed 11 Aug 2025
62. Environment-friendly polyurethane primer used universally by coiled materials and application thereof. SciSpace. https://scispace.com/papers/environment-friendly-polyurethane-primer-used-universally-by-s3nfcoypm8. Accessed 11 Aug 2025
63. Priming active siloxane composition, preparation method and application thereof. SciSpace. https://scispace.com/papers/priming-active-siloxane-composition-preparation-method-and-2lh7ja4jsx. Accessed 5 Aug 2025
64. Berger DM (1984) Zinc-rich primers: a survey of tests, specifications and materials. J Prot Coat Linings 1(2). https://scispace.com/papers/zinc-rich-primers-a-survey-of-tests-specifications-and-4r0ajkmynk
65. Preconstruction organic zinc primer and method for production and application. SciSpace. https://scispace.com/papers/preconstruction-organic-zinc-primer-and-method-for-4jk1zaj4ev. Accessed 5 Aug 2025
66. Liu W et al (2024) Intensified metallic effect and improved tribocorrosion resistance through microwave-based fabrication of metallic powder coatings. Prog Org Coat 189:108218. https://doi.org/10.1016/j.porgcoat.2024.108218
67. Van Bo Nguyen (2023) Optimal model-based control for automated robotized abrasive blasting system. https://scispace.com/papers/optimal-model-based-control-for-automated-robotized-abrasive-4s9up457w4. Accessed 5 Aug 2025
68. Zhang YZ et al (2014) Anti-corrosive epoxy primer modified with nano particles. Mater Sci Forum 789:112–116. https://doi.org/10.4028/WWW.SCIENTIFIC.NET/MSF.789.112
69. Self-healing polyurethane based on disulfide bond and hydrogen bond. ResearchGate. https://www.researchgate.net/publication/319415109_Self-healing_polyurethane_based_on_disulfide_bond_and_hydrogen_bond. Accessed 5 Aug 2025

70. Muha K, Hulme SP, Harakal ME (1997) Low odor amine catalysts for polyurethane flexible slabstock foams based on polyester polyols. US5591780A. https://patents.google.com/patent/US5591780A/en. Accessed 5 Aug 2025
71. Finch C (2004) Adhesion and adhesives technology—an introduction, 2nd edn. AV Pocius. Carl Hanser Gardener Verlag, Munchen, 2002. Polym Int 53(9):1394–1394. https://doi.org/10.1002/pi.1458
72. Jones FN et al (2017) Organic coatings: science and technology, 4th edn. Wiley. https://www.wiley.com/en-us/Organic+Coatings%3A+Science+and+Technology%2C+4th+Edition-p-9781119026891. Accessed 5 Aug 2025
73. Laser cleaning technologies in industry: mechanisms, applications, and future directions for sustainable surface treatment. SciSpace. https://scispace.com/papers/laser-cleaning-technologies-in-industry-mechanisms-otddwjw213jp. Accessed 5 Aug 2025

Chapter 3
Curing and Drying Processes

3.1 Curing and Drying of PU Coatings

As a versatile class of protective and functional materials, polyurethane (PU) coatings have seen widespread adoption across industries including automotive, marine, wood finishing, and construction, owing to their superior adhesion, flexibility, chemical resistance, and durability. The ultimate performance of these coatings is highly dependent on the curing and drying processes, which convert the applied liquid formulation into a solid, crosslinked film. This review presents a comprehensive synthesis of the fundamental mechanisms, coating types, influencing factors, evaluation methods, and recent advancements in PU coating curing and drying, drawing upon both established polymer engineering principles and contemporary research findings.

3.1.1 Drying vs. Curing

Although often used interchangeably, drying and curing represent distinct stages in the film formation of polyurethane (PU) coatings. Drying primarily entails the physical evaporation of solvents or water from the coating, resulting in initial film consolidation and a "dry-to-touch" state. This stage is influenced by environmental factors such as temperature, humidity, and airflow and can proceed at ambient temperatures as low as 0 °C in solvent-based systems [1]. In contrast, curing involves chemical crosslinking reactions that establish the polymer network, thereby enhancing mechanical properties such as hardness, tensile strength, and resistance to solvents, acids, and abrasion. While curing can overlap with the drying stage, it typically requires specific triggers such as moisture, heat, or ultraviolet (UV) radiation to progress to completion.

Z. Zubair et al., *Functional Polyurethane Coatings*, SpringerBriefs in Materials, https://doi.org/10.1007/978-3-032-12730-3_3

In waterborne polyurethane coatings, the drying stage precedes curing, requiring the removal of water through conventional dryers, microwave systems, or infrared (IR) lamps. Effective water removal is critical to prevent defects such as bubbling or incomplete crosslinking during subsequent curing [2]. Typical drying times for consumer-grade water-based polyurethane coatings range from approximately 20 min to reach a "dry-to-touch" state, with 2–4 h required before recoating. In contrast, oil-based formulations generally require 6–10 h to achieve comparable readiness for subsequent application [3]. Full curing of two-component polyurethane systems, however, can extend up to 14 days at 25 °C, with the coating attaining optimal mechanical and chemical properties only upon completion of this period [1].

3.1.2   Curing Mechanisms

The fundamental curing mechanism of polyurethane coatings is based on the reaction of isocyanate (-NCO) groups with active hydrogen-containing compounds, such as polyols (-OH), water, or amines, resulting in the formation of urethane or urea linkages. This polyaddition reaction generates a three-dimensional polymer network, which imparts the characteristic mechanical, chemical, and barrier properties of the coating.

Moisture-Cured Systems: In one-component, moisture-cured polyurethane systems, atmospheric humidity reacts with free –NCO groups to form amines, which subsequently react to generate urea linkages. This exothermic process is self-accelerating, with initial linear chain growth occurring within the first 24 h, progressing to a robust polyurea network by approximately 72 h [4]. Hydrogen bonding within the polymer matrix enhances cohesive strength, although encapsulation of unreacted –NCO groups can retard late-stage crosslinking reactions.

Two-Component (2K) Systems: Two-component polyurethane systems involve the mixing of a polyol base with a polyisocyanate hardener immediately prior to application. This reaction produces films that are both flexible and hard, exhibiting superior adhesion and chemical resistance. The pot life and the usable time after mixing typically ranges from 2 to 6 h, with a dry-to-touch state achieved in approximately 12 h. Catalysts, such as metal soaps or amines, are often employed to accelerate the curing process.

UV-Curable Systems: UV-cured waterborne polyurethane (UV-WPU) coatings incorporate acrylic double bonds that undergo free-radical polymerization upon UV irradiation. Photoinitiators generate radicals that crosslink the prepolymer with reactive diluents, restricting curing to the irradiated regions and enabling precise patterning. This curing approach enhances tensile strength and chemical resistance, although pre-drying is required to remove water before UV exposure [2].

Other Variants: Oxidative crosslinking occurs in urethane oils or alkyd-based polyurethane systems through reactions with atmospheric oxygen. In blocked isocyanate systems, commonly used in one-component paints, the isocyanate deblocks

Table 3.1 Drying and curing of different PU types used in various applications

Types of PU	Description	Drying time	Full curing time/ conditions	Applications	Ref
One-component (1K) moisture-cured	Reacts with moisture present in air	1–12 h	7–14 days at 25 °C, 50% humidity	Wood finishes, varnishes; low VOC options available	[4]
Two-component (2K)	Accelerated by catalyst	12 h	14 days; pot life 2–6 h	Automotive, marine; high durability	[1]
Waterborne (e.g., UV-WPU)	Aqueous dispersions; eco-friendly	20 min to 4 h; requires dryers	UV exposure for crosslinking; full cure rapid post-drying	Wood, plastics; low odor, fast recoat	[2]
Oil-based	Oxidative drying with oxygen	6–10 h	24–48 h	Traditional wood coatings; longer pot life	[3]
Blocked isocyanate (1K)	Heat-activated deblocking	Solvent evaporation at low temp	Cure at 120–200 °C	Industrial baking	[5]

at elevated temperatures (120–200 °C) to initiate curing. Sol-gel enhanced polyurethanes (SiPU) incorporate organosilanes to form hybrid networks, thereby improving thermal stability, withstanding temperatures up to 206 °C [1].

In specialized applications, such as pavement binders, curing involves the reaction of –NCO groups with substrate –OH groups and water, forming urea linkages that progressively enhance mechanical strength over time [4, 6] (Table 3.1).

Fast-drying formulations, particularly water-based polyurethanes, provide recoat times of 2–3 h at 70 °F, making them advantageous for applications requiring rapid turnaround and operational efficiency [6].

3.2   Factors Influencing Curing and Drying

The curing and drying processes of polyurethane coatings are influenced by numerous parameters, which directly affect film quality, performance, and application efficiency. Key factors are summarized below:

Moisture and humidity: Moisture-curing polyurethane (MCU) coatings are formulated from an excess of polyisocyanates and hydroxyl-containing polymers such as polyester resins, polyether resins, epoxy resins, hydroxyl-functional acrylic resins, and castor oil derivatives. The typical composition includes prepolymers, low water content solvents, pigments, dehydrating agents, additives, and catalysts. These coatings generally exhibit an NCO content of 10–15% and an NCO/OH molar ratio of approximately three or higher. One-component (1K) MCU coatings cure through reaction with atmospheric moisture, where –NCO groups react with water to form amines and carbon dioxide. The resulting amines subsequently react with additional –NCO groups, generating urethane and urethane-methacrylate linkages within the film. A notable fraction of –NCO groups remains unreacted, and based on the curing mechanism, MCU systems can be classified as either moisture-curing or latent-curing.

Moisture-curing polyurethane (MCU) coatings offer a wide range of applications, performing effectively at temperatures above 0 °C and relative humidity levels between 30 and 98%. They are easier to apply compared to two-component polyurethane systems and demonstrate excellent film-forming capabilities even under humid conditions. MCU coatings require minimal substrate preparation, exhibit high surface tolerance, and provide strong adhesion to a variety of substrates. They also deliver outstanding resistance to abrasion, corrosion, environmental pollutants, water, mechanical stress, and even atomic radiation. The drying and curing rates of MCU coatings are strongly influenced by relative humidity and temperature; lower temperatures slow drying due to reduced atmospheric moisture, whereas higher temperatures accelerate reactions of –NCO groups. Consequently, environmental conditions must be carefully considered during formulation design to ensure optimal coating performance [7].

Temperature: Elevated temperatures accelerate solvent evaporation and reaction kinetics in polyurethane coatings; however, excessive heat can compromise mechanical properties, potentially reducing hardness, flexibility, and overall film performance [8].

Additives: Additives are essential in polyurethane (PU) production to control reaction kinetics, modify processing conditions, and tailor final product properties. Common additives include catalysts, chain extenders, crosslinkers, fillers, moisture scavengers, and colorants. Catalysts are employed to accelerate reactions, lower curing and deblocking temperatures, and enable efficient formation of urethane linkages. Aliphatic and aromatic amines such as diaminobicyclooctane (DABCO), organometallic compounds (e.g., dibutyltin dilaurate, dibutyltin diacetate), and alkali metal salts of carboxylic acids or phenols (e.g., calcium, magnesium, strontium, barium salts of hexanoic, octanoic, naphthenic, and linolenic acids) are commonly used. For tertiary amines, catalytic activity is influenced by both basicity and steric hindrance; higher basicity enhances activity, whereas steric hindrance reduces it. Catalysis occurs via complex formation, where electrons from the nitrogen atom of the amine are donated to the electrophilic carbon of the isocyanate. Metal-based catalysts are generally preferred over tertiary amines due to their lower volatility and toxicity. Metals facilitate the isocyanate–hydroxyl reaction by forming a complex with both reactants, wherein the positively charged metal center interacts with the electron-rich oxygen atoms of isocyanates and hydroxyls, ultimately promoting urethane bond formation.

Difunctional low molecular weight diols such as ethylene glycol, 1,4-butanediol, 1,6-hexanediol, cyclohexane dimethanol, diamines, and hydroxyl-amines (e.g., diethanolamine, triethanolamine) are commonly used as chain extenders, whereas polyols with functionality ≥3 serve as crosslinkers. Because isocyanates are highly sensitive to moisture, moisture scavenger compounds that react preferentially with water are incorporated to prevent undesired side reactions. Examples include oxazolidine derivatives and zeolite-type molecular sieves [9].

Substrate and Application: Porous substrates improve adhesion through mechanical interlocking but can also absorb moisture, potentially delaying the drying process. Film thickness, typically ranging from 50 to 200 μm, influences diffusion pathways, with thicker layers prolonging solvent entrapment and slowing overall drying and curing [10].

Environmental and Process Factors: Proper airflow facilitates solvent evaporation, preventing accumulation within the film. In multistage automotive coating lines, it is critical to synchronize the curing of each layer to avoid intercoat adhesion failures [11]. Modern drying techniques, such as infrared (IR) or microwave-assisted systems, can reduce drying times by 50–70% compared to conventional methods [12].

3.3   Defects Due to Improper Drying and Curing

Improper drying and curing conditions can lead to defects that compromise both aesthetics and functional performance. Common issues include orange peel caused by uneven drying, cracking due to rapid solvent evaporation generating internal stress, and blushing resulting from moisture condensation under high-humidity conditions [13]. In waterborne polyurethane systems, incomplete coalescence can result in milky or hazy films. In UV-cured coatings, overcuring may lead to brittleness due to excessive crosslinking. For polyurea formulations, partial solvent or reactive diluent evaporation during curing can reduce final film thickness if the coating is not applied at the recommended mil levels [14]. Mitigation of such defects requires precise process control and the use of additives, including flow agents and defoamers, to ensure uniform film formation and minimize surface imperfections.

3.4   Methods to Accelerate or Control Curing and Drying Processes

The curing and drying phases frequently constitute bottlenecks in industrial polyurethane coating applications, directly affecting productivity, energy consumption, and final film quality. Accelerating these processes can minimize downtime in sectors such as automotive, flooring, and furniture, while precise process control helps prevent defects including bubbling, cracking, and uneven hardness. This review examines both established and emerging strategies to accelerate or regulate curing (chemical crosslinking) and drying (physical solvent or water evaporation) in PU coatings, encompassing environmental controls, additive incorporation, application techniques, and advanced technological interventions.

3.4.1 Environmental Control Methods

Environmental parameters exert a significant influence on the drying and curing kinetics of polyurethane coatings, providing simple and non-invasive means to accelerate these processes. Temperature elevation is a primary factor; for solvent-based PU systems, increasing ambient temperatures to 30–40 °C can reduce drying times by approximately 50%, by enhancing both solvent volatility and reaction rates [15]. In one-component moisture-cured (1K) PU systems, elevated temperatures accelerate isocyanate hydrolysis and urea formation, enhancing curing rates. However, excessive temperatures above 80 °C can induce yellowing or thermal degradation, compromising film appearance and performance [16]. Controlled post-application heating using infrared (IR) lamps or ovens accelerates oxidative curing in alkyd-urethane hybrid coatings by promoting oxygen diffusion into the film, which facilitates faster hydroperoxide decomposition and crosslinking [17].

Humidity control is equally critical in regulating polyurethane curing. While high relative humidity (>70% RH) can accelerate the curing of moisture-cured PUs, it may also cause CO_2 bubbling within the film. Maintaining moderate humidity levels of 40–50% RH, often via dehumidifiers, optimizes reaction rates while minimizing defect formation [18]. In waterborne polyurethane systems, lower ambient humidity accelerates water evaporation, reducing the dry-to-touch time from approximately 4 h to under 2 h at 70 °F [3]. Augmenting airflow through fans or forced-air systems enhances mass transfer, facilitating solvent removal and preventing skinning, which is a common issue in thicker polyurethane films [19]. In woodworking applications, circulating warm air approximately 74 °F using air conditioning and fans can reduce polyurethane drying times from several days to just a few h [18]. However, excessive ventilation can induce orange peel textures, highlighting the need for balanced airflow control. While these methods are generally cost-effective, their efficiency is substrate-dependent: porous woods absorb heat unevenly, whereas metals conduct heat more efficiently, affecting drying uniformity [20].

3.4.2 Additive-Based Methods

Additives offer chemical control over curing dynamics by either accelerating crosslinking or stabilizing formulations to prevent premature reactions. Catalysts are primary accelerators; for example, organotin compounds such as dibutyltin dilaurate (DBTDL) enhance urethane formation in two-component (2K) systems, reducing pot life to 1–2 h while enabling full cure within 24 h [21]. Amine catalysts, such as tertiary amines, provide milder acceleration in moisture-cured polyurethane systems, offering effective curing while avoiding the toxicity concerns associated with organotin compounds [22]. For oxidative drying, metallic driers such as cobalt or manganese soaps facilitate oxygen uptake, reducing surface dry times in urethane oils from 10 h to 6 h [23]. Recent environmentally friendly alternatives, such as manganese-based complexes, have been developed to replace cobalt catalysts, maintaining comparable curing efficiency while reducing toxicity and environmental impact [24].

Moisture scavengers are used to control unwanted hydration, enhancing storage stability and preventing defects in humid environments. For example, Additive TI, a reactive oxazolidine, offers short-term dehydration in both 1K and 2K polyurethane systems by neutralizing residual water, thereby reducing CO_2 bubble formation and film haze [25]. Additive OF provides long-term stabilization by chemically binding water, making it particularly suitable for the polyisocyanate component. Other moisture scavengers, such as siliporite (molecular sieves) or Incozol-2, adsorb residual water, thereby suppressing gas evolution in bubble-prone polyurethane systems [26]. Accelerators such as Additive 101, a solvent-based formulation for Carbothane series polyurethanes, enhance cure rates without affecting viscosity, making them particularly suitable for spray application processes [27].

Chain extenders and plasticizers play a key role in modulating polyurethane curing. Polyaspartic chain extenders in hybrid systems facilitate rapid ambient curing, achieving dry-to-touch within 30–60 min, while plasticizers, such as phthalates, regulate film flexibility during the drying process [22]. UV stabilizers, such as benzotriazoles and hindered amines, prevent photodegradation during extended curing periods, indirectly maintaining film quality. Fillers, including silica, and flame retardants improve thermal stability, while phosphite-based antioxidants inhibit oxidative side reactions that could compromise the coating's performance [28]. Proper additive dosage is critical, as over-addition can narrow processing windows or increase viscosity, negatively impacting application and film formation [29].

3.5 Curing Characterization Techniques

Assessing the extent of curing is essential for quality assurance in polyurethane coatings. Spectroscopic techniques, such as Fourier Transform Infrared (FTIR) spectroscopy, can monitor the disappearance of the –NCO absorption peak at 2270 cm^{-1}, providing a reliable indication of reaction completion [10]. Differential scanning calorimetry (DSC) is used to quantify the exothermic enthalpies associated with polyurethane curing, providing insights into reaction kinetics. Thermogravimetric Analysis (TGA) evaluates the thermal stability of the coating after curing, helping assess its resistance to decomposition under elevated temperatures [30]. Mechanical testing, including pendulum hardness measurements and tensile modulus evaluation via Dynamic Mechanical Analysis (DMA), is employed to monitor the evolution of film properties during and after polyurethane curing [10]. Solvent rub tests, such as MEK double rubs, are used to assess crosslinking density in polyurethane coatings. Typically, exceeding 200 rubs indicates that the coating has achieved full cure [31]. Advanced analytical techniques, such as confocal Raman microscopy, enable spatial mapping of cure gradients in thick polyurethane films, providing detailed insight into crosslinking uniformity throughout the coating.

References

1. Polyurethane coating—an overview. ScienceDirect Topics. https://www.sciencedirect.com/topics/engineering/polyurethane-coating. Accessed 10 Aug 2025
2. Dall Agnol L, Dias FTG, Ornaghi HL, Sangermano M, Bianchi O (2021) UV-curable waterborne polyurethane coatings: a state-of-the-art and recent advances review. Prog Org Coat 154:106156. https://doi.org/10.1016/j.porgcoat.2021.106156
3. Greenfield M. How long does polyurethane take to dry? Today's Homeowner. https://todayshomeowner.com/general/guides/how-long-does-polyurethane-take-to-dry/. Accessed 10 Aug 2025
4. Sun M et al (2021) Mechanism of polyurethane binder curing reaction and evaluation of polyurethane mixture properties. Coatings 11(12):1454. https://doi.org/10.3390/coatings11121454
5. Chattopadhyay DK, Raju KVSN (2007) Structural engineering of polyurethane coatings for high performance applications. Prog Polym Sci 32(3):352–418. https://doi.org/10.1016/j.progpolymsci.2006.05.003
6. Fast drying polyurethane. LumberJocks Woodworking Forum. https://www.lumberjocks.com/threads/fast-drying-polyurethane.303969/. Accessed 11 Aug 2025
7. What is ambient moisture curing polyurethane. https://doxuchem.com/what-is-ambient-moisture-curing-polyurethane/. Accessed 11 Aug 2025
8. Fahmina Zafar S et al (2022) Effect of curing temperature on mechanical strength, hydrophobicity, chemical and thermal stability of Cardanol-derived Resol-polyurethane films/coatings. Polym Eng Sci 62(7):2185–2196. https://doi.org/10.1002/pen.25999
9. Sharmin E, Zafar F (2012) Polyurethane: an introduction. In: Zafar F (ed) Polyurethane. InTech. https://doi.org/10.5772/51663
10. Wang T. Quantification of curing, hardness development, and degradation in epoxy and polyurethane coatings
11. Review of coating and curing processes: evaluation in automotive industry. Physics of Fluids, AIP Publishing. https://pubs.aip.org/aip/pof/article/34/10/101301/2846660/Review-of-coating-and-curing-processes-Evaluation. Accessed 11 Aug 2025
12. Polyurethane coating drier alternatives for fast handling of coated wood items: a comprehensive review. BDMAEE. https://www.bdmaee.net/polyurethane-coating-drier-alternatives-for-fast-handling-of-coated-wood-items-a-comprehensive-review/. Accessed 11 Aug 2025
13. Chemical and mechanistic aspects of wood finishing: a review encompassing paints and clear coats. BioResources. https://bioresources.cnr.ncsu.edu/resources/chemical-and-mechanistic-aspects-of-wood-finishing-a-review-encompassing-paints-and-clear-coats/. Accessed 11 Aug 2025
14. 5 problems with polyurea and polyaspartic concrete floor coatings—southwest exteriors blog. https://www.southwestexteriors.com/blog/2021/november/5-problems-with-polyurea-and-polyaspartic-concre/. Accessed 10 Aug 2025
15. Polyurethane drying time guide. 24 hour floor. https://24hourfloor.com/polyurethane-drying-time-guide/. Accessed 11 Aug 2025
16. Speed up curing time by applying heat?. Woodworking Talk. https://www.woodworkingtalk.com/threads/speed-up-curing-time-by-applying-heat.123617/. Accessed 11 Aug 2025
17. Fristik B. The curing process. Waterlox Coatings Corporation. https://waterlox.com/guide-the-curing-process/. Accessed 11 Aug 2025
18. Accelerating polyurethane dry time. Wood Talk Online. https://www.woodtalkonline.com/topic/19766-accelerating-polyurethane-dry-time/. Accessed 11 Aug 2025

19. Best tips to speed up paint drying and curing. Coatings Directory. https://coatingsdirectory. com/blog/tips-to-speed-up-paint-drying-and-curing/. Accessed 11 Aug 2025
20. Curing methods. https://p2infohouse.org/ref/01/00777/curing.htm. Accessed 11 Aug 2025
21. Polyurethane coating—an overview. ScienceDirect Topics. https://www.sciencedirect.com/ topics/engineering/polyurethane-coating. Accessed 11 Aug 2025
22. What additives are needed for polyurethane adhesives? LinkedIn. https://www.linkedin.com/ pulse/what-additives-needed-polyurethane-adhesives-anna-wu-hqv1c/. Accessed 11 Aug 2025
23. Additive B. ANTI-SKINNING AGENTS
24. The curing process. Waterlox Coatings Corporation. https://waterlox.com/guide-the-curing-process/. Accessed 11 Aug 2025
25. Moisture scavenger for reduced defects in polyurethanes. Borchers: A Milliken Brand. https:// borchers.com/news-events-blog/how-moisture-scavengers-reduce-defects-in-polyurethane-coatings/. Accessed 11 Aug 2025
26. What additive could I use to suppress the high amount of gas bubbles in my polyurethane system? ResearchGate. https://www.researchgate.net/post/What_additive_could_I_use_to_supress_the_high_amount_of_gas_bubbles_in_my_polyurethane_system. Accessed 11 Aug 2025
27. Additive 101. Carboline. https://www.carboline.com/products/product-details/Additive-101/. Accessed 11 Aug 2025
28. Common additives for polyurethane industry. Imenpol blog. https://imenpol.com/blog/en/educational/common-additives-for-polyurethane-industry/. Accessed 11 Aug 2025
29. Polyurethane additives: an overview guide. https://www.cficarbonproducts.com/polyurethane-additives/. Accessed 11 Aug 2025
30. Study of water resistance of polyurethane coatings based on microanalytical methods. https:// www.mdpi.com/2073-4360/16/24/3529. Accessed 11 Aug 2025
31. Is polyurethane supposed to be this soft? r/woodworking. https://www.reddit.com/r/woodworking/comments/1bzug7z/is_polyurethane_supposed_to_be_this_soft/. Accessed 11 Aug 2025

Chapter 4
Performance and Properties

4.1 Introduction

Polyurethane coatings are renowned for excellent wear resistance, chemical resistance, mechanical properties, and corrosion resistance among various types of coatings. They have balanced mechanical properties, moderate to high hardness, good flexibility, and impact resistance arising from its soft and hard structural components. PU is a block copolymer with alternating hard segments (small diols and isocyanate-derived carbamates) and soft segments (long-chain polyols) shown in Fig. 4.1. These segments are thermodynamically incompatible and phase-separate at the nanoscale [1, 2]. Among these segments, the hard segment acts as physical crosslinkers, contributing tensile strength, hardness, and tear resistance, while soft segment imparts flexibility and elongation to the PU performance [3]. The overall mechanical performance of PU coatings depends on a complex interplay between their chemical structure, the ratio of hard and soft segments, crosslink density, and the use of additives or nanofillers [4]. Recent research has focused on tailoring these parameters to enhance specific properties such as hardness, flexibility, and impact resistance. Understanding these properties and how to improve them are pivotal for developing coatings capable of meeting challenging environmental and operational demands.

4.2 Hardness

In the context of polyurethane coating, hardness is a measure of a coating's resistance to plastic deformation, scratching, and wear. Polyurethane coatings exhibit a wide range of hardness, from very soft to very hard, typically ranging from 25 Shore OO to 75 Shore D, according to Gallagher Corporation. Many polyurethane

Z. Zubair et al., *Functional Polyurethane Coatings*, SpringerBriefs in Materials,
https://doi.org/10.1007/978-3-032-12730-3_4

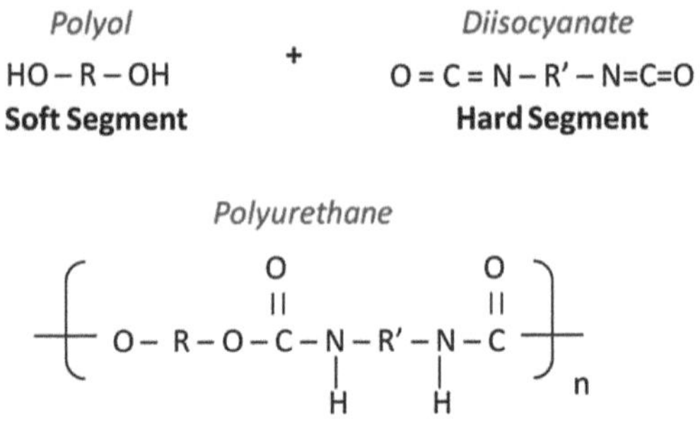

Fig. 4.1 Chemical structure of polyurethane containing hard and soft segments, their ratio decides the overall properties of polyurethane [5]

Polyol + *Diisocyanate*

HO – R – OH   O = C = N – R′ – N=C=O

Soft Segment   **Hard Segment**

Polyurethane

$$\left[\!\!-\!\! \begin{array}{c} O - R - O - \overset{\displaystyle O}{\overset{\displaystyle \|}{C}} - \underset{\underset{H}{|}}{N} - R' - \underset{\underset{H}{|}}{N} - \overset{\displaystyle O}{\overset{\displaystyle \|}{C}} \!\!-\!\! \end{array} \right]_n$$

coatings fall within the range of 55 Shore A to 75 Shore D, which is often considered the sweet spot for achieving a balance of performance properties like abrasion resistance, load-bearing capacity, and flexibility. A study by Samad et al. measured the energy attenuation coefficient (α_E) for different hardness levels of neat polyurethane which characterizes how much acoustic energy is lost per unit length (measured in nepers per centimeter, Np/cm). Their results show that the harder, denser PU exhibits higher attenuation, meaning it dissipates more acoustic energy per unit length than softer PU [6]. The hardness of PU coating is decided based on the end application and environmental exposure, especially where mechanical durability, abrasion resistance, and scratch resistance are critical, for examples in floor coatings, automotive coatings, wood and furniture and industrial tools, etc. So, different mechanisms are applied to enhance its hardness; increase the hard segment content, incorporation of rigid nanofillers, hybridization with epoxy resin, silane additives, and introduce crosslinking are some of them [7, 8]. Following are some strategies to increase the hardness of the PU that eventually increase the hardness of PU coating.

4.2.1 Incorporation of Rigid Nanofillers

Introducing nanofillers into polyurethane (PU) particularly for coatings can significantly improve mechanical performance, hardness, etc. Different studies use different name of nanofiller that enhance the hardness and mechanical properties of polyurethane that eventually increase the PU coating hardness. A study by Liang et al. [9] demonstrates that graphene oxide grafting causes the 15.9% times increase the hardness of pure PU. Another study incorporates the multi-walled carbon nanotubes (MWCNT), two types of silica nanoparticles, and MWCNT/silica as hybrid fillers in PU [10]. When different nanofillers are well-dispersed and surface-functionalized in PU, they bond strongly with PU's urethane network, increasing crosslink density. The result is a stiffer, harder coating surface. CNC-PU composites cured at 150 °C reached 3H pencil hardness and Shore D81, significantly harder than typical PU coatings (Shore D20–60). The high hardness was attributed to the stiffness and crystallinity of CNCs and dense urethane linkage formation [11]. Graphene nanoplatelets, cellulose nanocrystals (CNC), nano silica, silica/Mg (OH)$_2$

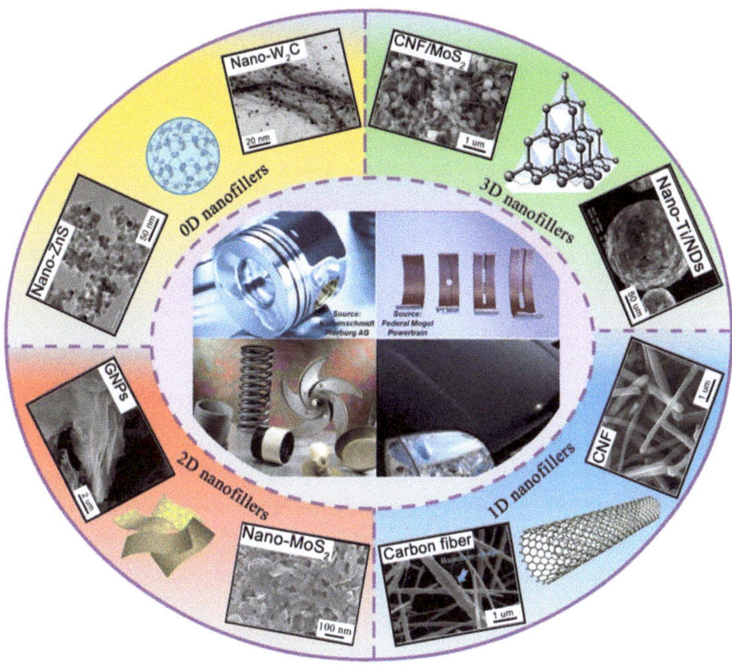

Fig. 4.2 Some valuable nanofillers that use as reinforcement in PU coatings [13]

hybrid fillers, and halloysite nanotubes are valuable nanofillers that have been incorporated into polyurethane (PU) coatings to enhance hardness and mechanical properties, as demonstrated by experimental studies conducted by various authors [12–14]. Some nanofillers used in PU coatings are shown in Fig. 4.2.

4.2.2   Hybridization with Epoxy Resin

Combining polyurethane (PU) with epoxy resin, i.e., creating PU/epoxy hybrid coatings has been shown in recent studies to significantly enhance key mechanical properties, especially hardness, through additional crosslinking and increased network rigidity. A study by Bekbayeva et al. increased the mechanical properties of PU coating by incorporating epoxy resin; with 5% epoxy, tensile strength rose from 53.5 MPa to 68 MPa and hardness increased from Shore A 81. With 15% epoxy, tensile strength reached 86.3 MPa and hardness peaked at Shore A 98 [15]. Other studies also mentioned that the same hydroxyl-epoxy crosslinking boosts adhesion and shortens drying time due to more reactive groups, because epoxy introduces reactive epoxy groups that react with PU–polyol/hydroxyl species, forming a densely crosslinked network shown in Fig. 4.3. This increases hardness (Shore A), tensile strength, and adhesion, although elongation typically decreases [16, 17].

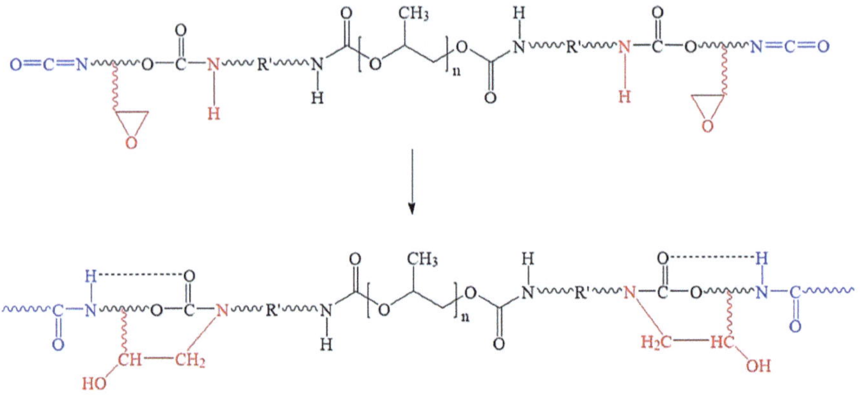

Fig. 4.3 Bonding representation between PU and epoxy, showing how epoxy incorporation increases active sites of PU and the crosslinking density of overall hybride PU/epoxy resin coating [18]

4.2.3 Addition of Silane Additives

Silane coupling agents (SCAs) are organosilicon compounds used widely to improve the overall hardness and mechanical properties of the PU coatings. Recent studies demonstrate that the addition of different types of silane agent significantly increases the performance of PU coating in high-end applications. Elmas Kant et al. [8] used TMSCH (Trimethoxysilyl-propylcarbamoyloxyhexane), a silane coupling agent in PEG-600/IPDI-based PU. Lu et al. [19] use CTS (organofunctional silane coupling agent) in PU/inorganic composites. Some other studies use silane-coupled sodium silicate (KH-560 treated) and amino-functional siloxane silane (APTS) in PU coatings [20]. Their results exhibit that addition of silane coupling agents dramatically improves the hardness, scratch resistance, wettability, and flexural modulus of added PU compared to neat PU. Because addition of silane coupling agent forms additional urethane bonds via reaction of the silane's carbamoyloxy and isocyanate groups that leads to increase the crosslinking density, acts as hard segment crosslinker, and creates Si–O–Si network with PU chains [21].

4.3 Flexibility

Flexibility is one of the hallmark characteristics of polyurethane (PU) coatings, enabling their extensive use across industries such as construction, automotive, electronics, biomedical, and marine sectors. Flexibility refers to a PU coating's ability to deform (e.g., bend, elongate, or stretch) without cracking or losing adhesion. It depends on the soft segment content, phase morphology, crosslink density, and segment chemistry. Flexible PU coatings offer significant advantages like the ability to withstand dynamic loading, deformation, substrate movement, thermal cycling, and other mechanical stresses without cracking or delaminating [12].

As previously mentioned, the segmental architecture of PU includes soft and hard segmet. The soft segment imparts flexibility and elasticity, while the hard segment provides rigidity and strength. Their ratios determined the nature of end properties of PU coatings. The flexibility of PU generally characterized by parameters such as elongation at break and tensile strength [22]. Recent studies have explored strategies to optimize flexibility while maintaining or even improving mechanical and chemical resistance, including the use of hybrid materials, nanofillers, and reactive diluent tuning.

A latest study indicates that the elongation at break for modern PU coatings and hybrid composites can range between 49 and 98%. Zhang et al. developed a non-isocyanate polyurethane (NIPU) system that achieved ceramic-like hardness and polymer-like flexibility by introducing epoxy-oligosiloxane nanoclusters into a flexible backbone. The resulting coating exhibited high surface hardness (5H–7H) while remaining highly bendable (2 mm bending diameter), demonstrating a key balance between rigid and flexible domains [23]. Liang et al. synthesized graphene oxide (GO)-grafted waterborne polyurethane (WPU) coatings, this demonstrates that graphene and other nanomaterials can be used strategically to reinforce coatings without sacrificing flexibility when proper dispersion and compatibility are ensured. A recent study by Akin et al. investigated the press formability of UV-cured polyurethane acrylate (PUA) films by varying the type of diluent. Monofunctional acrylates produced more flexible films due to reduced crosslink density. Multifunctional acrylates led to brittle films prone to cracking under strain. When a monofunctional reactive diluent was used, flexibility was excellent, with no necking or cracking observed [24].

Incorporating nanofillers such as graphene oxide, silica, zinc oxide (ZnO), and layered double hydroxide (LDH) is a common strategy to enhance mechanical performance, including flexibility [25]. At optimal concentrations, nanofillers interact synergistically with the polymer chains, increasing flexibility and toughness without compromising other properties. Flexible PU foams reinforced with graphene and boron nitride exhibited increased elongation and toughness compared to neat PU foams [26]. Bio-inspired modifications, such as osthole-infused flexible PU coatings, have been shown to maintain flexibility while introducing self-healing and antifouling properties [27]. A study by Xie et al. developed relation that an optimal diluent content increases both tensile strength and flexibility; however, exceeding this range can lead to decreased performance or embrittlement. Positron annihilation lifetime spectroscopy (PALS) studies revealed a strong correlation between free volume parameters and elongation at rupture; increased free volume typically leads to increased flexibility [28].

4.4 Impact Resistance

Impact resistance in PU coatings is its ability to absorb and dissipate kinetic energy from sudden mechanical shocks, thus preventing cracking, delamination, or substrate damage. This property is vital in sectors such as automotive, aerospace, civil construction, and consumer electronics, where coatings must withstand repeated impacts, abrasion, and sometimes even erosive environments. Recent research has significantly advanced the understanding and enhancement of the impact resistance of PU coatings, focusing on both the modification of molecular structures and the incorporation of functional additives. In a 2024 study, PU coatings reinforced with 0–2 wt.% LDH nanoparticles were evaluated via a steel ball impact test following ASTM D-3029. Impact energy increased from 5.3 J (pure PU) to 15.9 J at 1.5% LDH [25]. Guo et al. showed that spray-on polyurea coatings applied to aluminum sheets offer superior impact protection and toughness, especially effective under high-strain, ballistic loading conditions and can enhance resistance more effectively than simply increasing panel thickness. Phase-segregated polyureas, due to their elastomeric/glassy network, offer high energy dissipation, which is instructive for advanced PU formulations [29]. Liang et al. grafted graphene oxide (GO) chemically into waterborne PU coatings. At 0.7 wt.% GO, tensile strength increased 65%, with pendulum hardness and abrasion resistance improved by 16% and 28%, respectively [9]. Figure 4.4 explains how the carbon nanofillers increase the impact resistance of the PU coatings.

4.5 Chemical Resistance and Durability of Polyurethane Coatings

Polyurethane coatings are widely used in protective applications due to their mechanical robustness, adhesion, and flexibility. Recently, significant research has focused on enhancing PU's chemical resistance and overall durability, addressing challenges such as solvent attack, acid/base exposure, weathering, and barrier

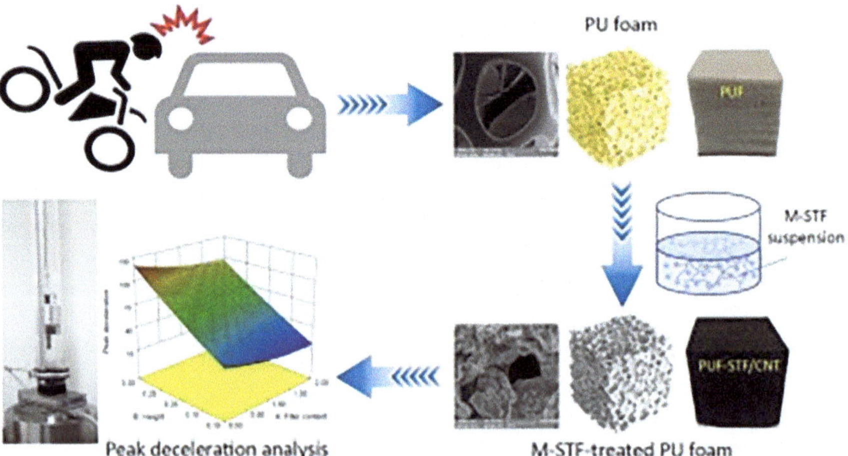

Fig. 4.4 Enhancement of impact resistance by addition of carbon nanofillers in PU foam [30]

performance. Recent research underscores not only the versatility of PU chemistry but also innovative approaches to further improve their properties through molecular engineering, hybridization with other polymers, and the introduction of nanotechnology and sustainable, bio-based component.

4.5.1   Chemical Resistance of PU Coatings

PU coating's resistance depends on several factors such as backbone chemistry, crosslinking density, and hydrophobicity of PU coating. Generally, high crosslinking density leads to less permeable and more stable coatings. Polyurethane coatings are generally resistant to caustic chemicals, acids, alkalis, hydraulic oils, and many organic solvents. However, their performance can vary with temperature, chemical concentration, and substrate nature [31]. Recent studies explained the different mechanism to improve the chemical resistance of PU coatings.

Research on PU/epoxy hybrid coatings (PUAE series) reported that intermolecular hydrogen bonding between epoxy hydroxyl groups and PU isocyanate groups enhances crosslink density. This densified network reduces free volume and effectively minimizes penetration paths for corrosive species, thereby improving both mechanical and corrosion resistance, especially at 15 wt% epoxy content [15]. Fluorinated polyurethane (FPU) coatings, containing fluorinated blocks and layered pigments such as mica flakes or aluminum tripolyphosphate, exhibited excellent chemical stability, low surface energy, and superior salt spray resistance [32]. Research on silane-coupled inorganic fillers indicates that incorporating silane-functionalized ceramic nanoparticles strengthens the PU network via covalent bonding, improving hydrophobicity, reducing micro-cracks, and limiting moisture/chemical exposure [33].

Incorporating acrylics in PU matrices fosters synergistic effects, enhancing chemical and corrosion resistance significantly. Notably, pure PU shows strong resistance to aqueous and alkali exposures, while hybrid PU/acrylic (10% AK content) demonstrated the greatest resistance across tested chemicals outperforming both neat PU and higher AK blends [34]. The drive toward sustainability has resulted in isocyanate-free PU (NIPU) synthesized from vegetable oils, which have shown remarkable resistance to acids and water, though some sensitivity to alkalis remains in certain formulations, largely due to hydrolysable ester linkages [35]. Other studies have incorporated nanomaterials such as SiO_2–ZnO core-shell particles or MXene sheets into PU matrices, enhancing passive barrier properties and introducing self-healing features. For instance, MXene/PU composites displayed superior anticorrosion performance under aggressive corrosive conditions, attributed to improved barrier function and active/passive corrosion inhibition [36]. Chemical resistance in PU coatings emerges from sophisticated molecular design, embedding nanofillers (especially GO), hybridizing with epoxy systems, incorporating fluorinated segments, or using silane-inorganic precursors. These modifications improve network density, surface hydrophobicity, and barrier morphology, collectively blocking aggressive media and extending durability.

4.5.2   Durability and Long-Term Performance of PU Coatings

Durability in coatings covers performance under mechanical stress (abrasion, impact, flexing), as well as resistance to UV, temperature changes, and environmental aging. PU is prized for their balance of rigidity and flexibility, which can be tailored via the selection of monomeric polyols and isocyanates. Combined degradation models like those of Gupta et al. enabled prediction of gloss loss, surface roughening, and degradation in fracture toughness, offering a comprehensive durability assessment over time under UV, chemical, and abrasion stresses [37]. Nanofillers such as graphene oxide or carbon-based materials effectively reduce permeability to moisture and corrosive ions, and slow chemical ingress. Nazari et al. note that carbon-based nanomaterials improve barrier performance in PU coatings significantly, limiting diffusion pathways and enhancing long-term chemical durability.

PU coatings show notable stability under heat and cycling; addition of aromatic structures or hybrid fillers can raise the glass transition and decomposition temperatures, thereby extending their usable range. Exposure to sand, wind-blown particles, and temperature extremes can cause microcracking and surface roughening, leading to eventual loss of protection. However, advanced PU formulations, especially those with graphene oxide or silica nanoparticle additives, show better long-term maintenance of smoothness, adhesion, and barrier properties [38, 39].

4.5.3   Corrosion Resistance

A core application for PU coatings is corrosion prevention. Recent studies confirm that, electrochemical impedance spectroscopy (EIS) and accelerated salt spray tests show enhanced resistance in both traditional and modified PUs. Graphene, halloysite nanotube, and other nanofiller enhancements further reduce permeability and enable controlled release of inhibitors. In a study, waterborne Jatropha oil-based polyurethane coatings on steel maintained high impedance modulus (up to 10^7 Ω cm^2) after extended salt water immersion, far exceeding many conventional coatings and indicating outstanding anticorrosion capability [40].

4.5.4   Advancement in Durability of PU Coatings

Nano-Composites: Inclusion of graphene oxide or nano-SiO2 into PU matrices improves both chemical resistance and durability, owing to increased crosslink density and enhanced barrier effects. This also translates into better resistance to erosion, abrasion, and environmental aging [41].

Bio-based and Sustainable Innovations: Recent studies by Li et al. [40] showed that NIPUs from soybean, linseed, and jatropha oils offer competitive or even superior resistance compared to conventional petrochemical PU, particularly in

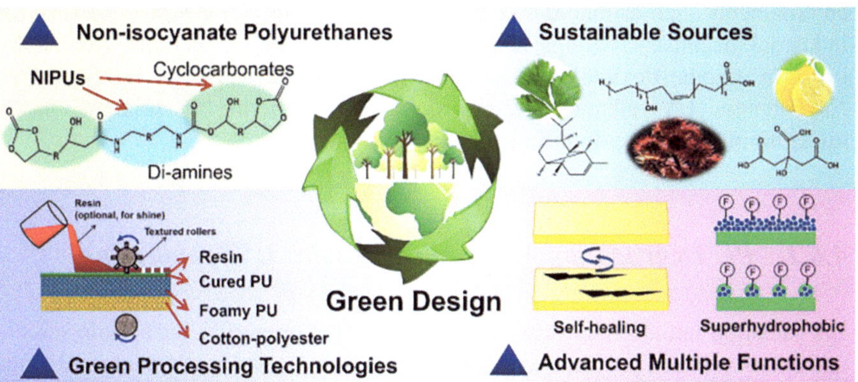

Fig. 4.5 Elaboration of green design, including the use of sustainable sources and eco-friendly processing methods for polyurethane (PU), which exhibits advanced functionalities such as self-healing and super hydrophobicity [37]

corrosion and water/solvent resistance, while maintaining mechanical flexibility and processability.

Multifunctionality (Self-healing, Antibacterial, and Smart Coatings): Self-healing PU coatings with dynamic covalent (disulfide, Diels–Alder) bonds or boronated-crosslinked networks allow for microdamage remediation, greatly enhancing coating lifetime and durability under physical and chemical stress show in Fig. 4.5 [41]. New PU coatings that incorporate antibacterial agents or bio-based antimicrobial networks are being developed for applications in medical devices, food packaging, and more [35].

Despite substantial advances, continued research into long-term durability, field performance under real-world cycling/aging, and broader adoption of green chemistry routes remains crucial for further progress.

4.6 Weathering and UV Resistance of PU Coatings

Polyurethane coatings are extensively used for outdoor protection due to their mechanical and aesthetic performance. However, UV radiation and weathering degrade PU's chemical and physical integrity, leading to discoloration, cracking, loss of gloss, and adhesion failure. Understanding weathering and UV resistance mechanisms in PU coatings and advancements to mitigate such effects remain critical for extending the service life of PU coated products. Accelerated weathering of bio-based PU coatings revealed that higher rigid-segment (hard segment) content improves resistance to UV-induced cracking and degradation. In formulations with molar ratio [NCO]/[OH] = 1.2 (vs 0.8), coatings sustained longer exposure with fewer cracks and greater retention of thermal stability [42]. The versatility, durability, and high-performance qualities of PU make them widely used in outdoor environments such as automotive body finishes, architectural protective surfaces, industrial machinery coatings, and marine and aerospace applications. Their ability to form tough, flexible, and aesthetically pleasing films is highly valued. Nevertheless, UV radiation and environmental weathering are principal factors undermining their long-term functionality [43].

4.6.1   PU Degradation Mechanisms Under UV/Weathering

UV radiation initiates photochemical reactions that primarily cause chain scission and crosslinking in polyurethane polymers. The energy from UV photons disrupts chemical bonds, particularly scission of urethane linkages and aromatic groups, oxidation of methylene groups and amide functional groups, formation of hydroperoxides and free radicals, etc. Additionally, weathering involves combined stressors such as moisture, temperature cycling, wind, and pollutants which accelerate oxidation and hydrolytic degradation, causing loss of mechanical strength and gloss [44]. Followings are some ways how the UV and weathering cause the losses of multiple properties of the PU coatings.

4.6.2   Photooxidation (Chain Scission and Chemical Modification)

Photodegradation in PU coatings originates primarily from UV radiation interacting with aromatic structures (e.g., aromatic isocyanates) and urethane/urea bonds, triggering photooxidation and chain scission. Photofries rearrangements, oxidation of ether linkages in soft segments, and cleavage of hard segment urea groups lead to brittleness and surface defects [45]. Experiments on photoaged PU microplastics found yellowing due to oxidation of aromatic isocyanate residues to quinone-imide structures, along with photofries rearrangement of urethane linkages. FTIR analysis revealed increased carbonyl index (peak at ~1704 cm^{-1}), broad OH stretching (~3200–3600 cm^{-1}), and quinonic C=C (~1597 cm^{-1}) bands—clear signs of photooxidation and chain degradation [46]. Photooxidation initiates via free-radical generation, often triggered by UV-absorbing chromophores or impurity groups like hydroperoxides. These radicals react with oxygen to form peroxides, hydroperoxides, and carbonyl-containing products. Subsequent decomposition (e.g., Norrish type I/II) leads to chain scission and material brittleness. The autocatalytic nature accelerates degradation with time [47].

Accelerated UV exposure causes nano- and micropore formation, leading to capillary networks that eventually reach the substrate, enabling ingress of moisture and corrosives. Gradually, cracks, craters, macrospores, and fractures around pigments/fillers emerge, accelerating breakdown of physical properties like adhesion and barrier performance [48]. Polymers (PU/PU-reactive coatings) exposed to UV and salt spray showed significant surface roughness, adhesion loss, and increased water contact angle changes which are shown in Fig. 4.6. FTIR and TGA analyses confirmed oxidative structural changes depending on aging conditions (UV vs NaCl solution vs thermal). Adhesion decreased most under NaCl aging [49]. Natural outdoor exposures, e.g., in marine or tropical environments accelerate these phenomena due to solar UV, salt spray, and thermal cycling.

4.6.3   Effects of UV and Weathering on PU Coatings

Photodegradation of PU under UV light primarily involves photooxidation, including Norrish I and II reactions, leading to chain scission of urethane groups, formation of radicals, and increased carbonyl content. A study by Acierno et al. reported that UV radiation caused urethane bond cleavage, increasing carboxylic acid and aldehyde groups in waterborne PU systems [50]. FTIR analysis by Monteiro et al. revealed a progressive reduction in the amide II band (1530 cm^{-1}) and an increase in carbonyl content (1706 cm^{-1}) during UV exposure, indicating cleavage of urethane linkages and oxidation products. Random chain scission (Norrish I) dominates in aliphatic PU,

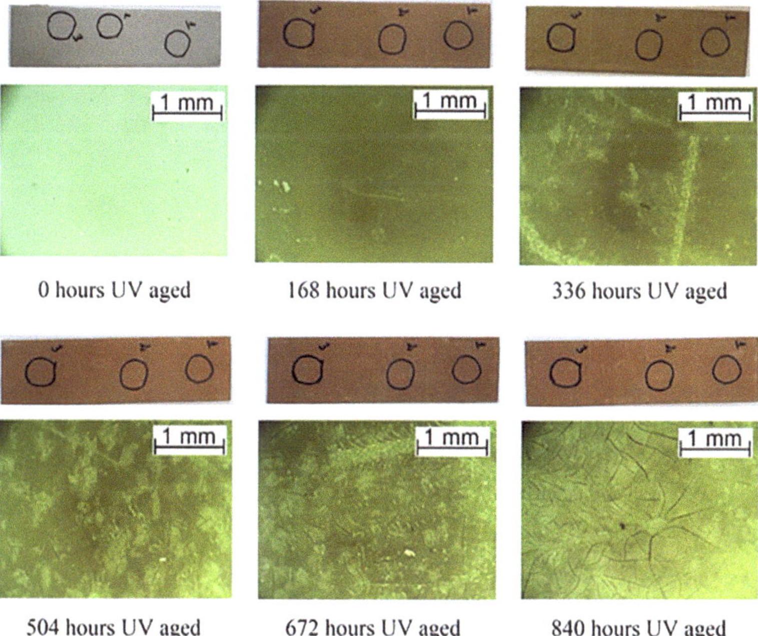

Fig. 4.6 Microscopic analysis of surface degradation of PU coatings, induced by UV aging at different hours [49]

generating radicals that propagate degradation and sometimes crosslinking [33]. One example of UV/O$_2$ induced reaction of PU in Fig. 4.7.

Prolonged exposure of UV light and weathering results in discoloration and yellowing due to chromophore alterations, loss of gloss and surface roughening, cracking and embrittlement, and adhesion loss leading to delamination (Fig. 4.8). These

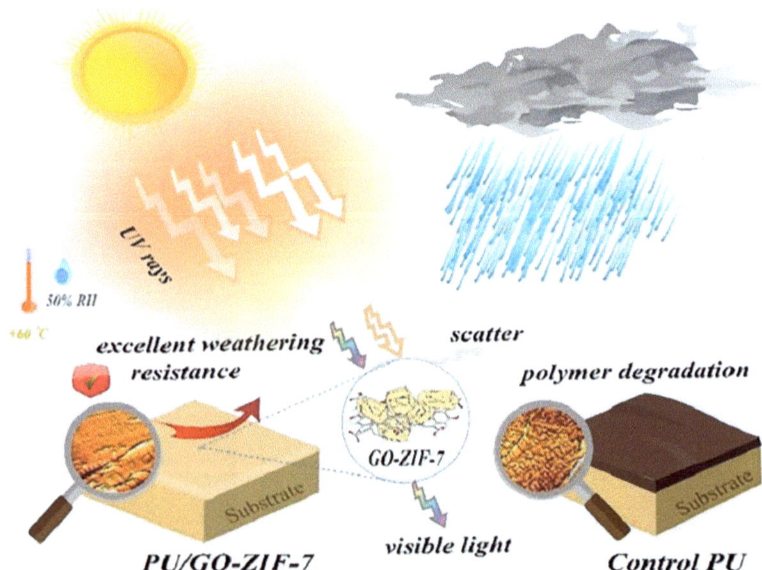

Fig. 4.7 Schematic depiction of the UV/O$_2$-mediated conversion of para-phenylene bis-urea linkages in a polymer into azobenzene-type −N=C−N= chromophore units [51]

Fig. 4.8 Representing how weathering and UV exposure degrade the polymer surface and caused discoloring of PU coating. In comparison coating with ZIF treated shows good weathering resistance [44]

phenomena significantly reduce the coating's protective effect and visual appeal [44]. In polyurea (related PU family), discoloration and crazing appeared within 3 weeks and worsened over time. Photodegradation increases brittleness and elastic modulus, decreases tensile strength, changes friction behavior, and deteriorates mechanical resistance [52].

A study exposed PU coatings on composites to 200 days of cyclic environmental testing (salt, temperature, humidity) [53]. Observed effects included:

- Gloss drop from 93 GU to 78 GU
- Contact angle reduction from ~92° to 70°
- Adhesion strength loss from 11.3 to 6.8 MPa
- Wear depth increased by over 350%
- Thermal degradation onset (5 wt.%) lowered by 11 °C

Recent high-quality studies consistently show that UV radiation alone causes photooxidative degradation, while weathering (combined UV, moisture, and thermal cycling) significantly accelerates both chemical breakdown and physical failure in PU coatings. Effective UV stabilizers, hydrolysis inhibitors, and multilayer or hybrid coatings are therefore essential for extending outdoor PU coating service life.

4.6.4   Strategies to Improve UV Resistance and Weathering Durability

As we discussed earlier that the exposure of UV and weathering leads to surface cracking, yellowing, loss of gloss, reduced mechanical strength, and eventual coating failure. Additionally, atmospheric oxygen, moisture, and pollutants accelerate this degradation via hydrolysis and oxidation pathways. Recent strategies go beyond traditional stabilizer additives and now include molecular architecture modification, nanocomposite engineering, and bio-derived photo stabilizers [54]. Several advanced studies ranging from UV absorber integration and nanoparticle reinforcement to bio-based hybrid systems and smart polymer design. These techniques aim to shield PU materials against photodegradation, retain their visual and functional properties over long-term outdoor exposure, and align with the demands of next-generation, high-durability coatings.

4.6.4.1   Incorporation of UV Absorbers and Light Stabilizers

UV absorber and stabilizers are types of additives that incorporate into materials to save them from attack of UV and degradation caused by UV exposure. UV absorbers function by absorbing harmful UV radiation and dissipating it as harmless heat, preventing initiation of photodegradation reactions and quenching of excited states in the PU matrix [55].

UV absorbers capture UV photons (typically 280–400 nm), converting that energy into harmless heat which is dissipated into the material matrix. This prevents photons from triggering polymer degradation. And some other UV stabilizer work by stopping photons from initiating chain reactions, UV absorbers help limit photo-oxidative free-radical formation that otherwise breaks polymer bonds. Although HALS are not true absorbers, they neutralize polymer radicals in a cyclic regeneration mechanism (the Denisov cycle), complementing UV absorbers. The combination significantly improves durability. Certain additives (e.g., metal complexes) can deactivate excited polymer chromophores directly, returning them safely to the ground state without producing singlet oxygen. This quenching hides in tandem with absorption mechanisms [56].

The UV stabilizers can incorporate within the polymer by two ways: chemical grafting within the PU backbone chain and synergistic use of UV absorbers. BZT-type absorbers such as Chiguard R-455 (a benzotriazole derivative) can be covalently incorporated in situ into the polymer backbone, forming permanent non-migrating anti-UV structures while maintaining colorless transparency and mechanical performance [43]. Another study incorporates the methoxybenzophenone moieties into the backbone chain of the polyurethane, the results show that incorporation of additives into the polymer increases the maximum decomposition temperature and the amount of char residue and show good resistance to UV radiation, with small changes in the mechanical properties and surface morphology [57]. Combining hydroxyphenyl-benzotriazole or triazine UV absorbers (Tinuvin 328, 479, 384, 292) with Hindered Amine Light Stabilizers (HALS) provides excellent protection, UV absorbers screen damaging rays while HALS capture the radicals formed, regenerating via the denisov cycle and continuously protecting the PU [58]. Benzophenones, benzotriazoles, triazines (hydroxyphenyl-triazines), and salicylates are some types of organic UV absorbers that commonly used in the PU coatings. Some hybrid systems that use as UV absorbers and stabilizers are hybrid organic microfine particles, higher molecular-weight grafted absorbers, Schiff-bases and their organometallic complexes, and natural UV-active compounds [54, 59]. Mechanism of UV absorber is shown in Fig. 4.9.

Nanoparticle-Based UV Shielding

Nanoparticles act as UV reflectors or absorbers and can also form a physical barrier, increasing the tortuosity for oxygen, water, and UV radiation diffusion. Many inorganic nanoparticles (TiO_2, ZnO) absorb UV radiation by promoting electrons across a wide band-gap, converting photon energy into heat and preventing it from reaching underlying material. In other way, they scatter and reflect UV light, especially when particle size is comparable to UV wavelengths. Well-dispersed Nano-ZnO or TiO_2 can scatter effectively without compromising visible transparency [61]. Several nanoparticles like Titanium Dioxide (TiO_2), Zinc Oxide (ZnO), silver nanoparticles, Metal–Organic Frameworks (MOFs), hybrid nanoparticles (TiO_2–SiO_2 mesoporous rods), and graphene oxide are some common types of

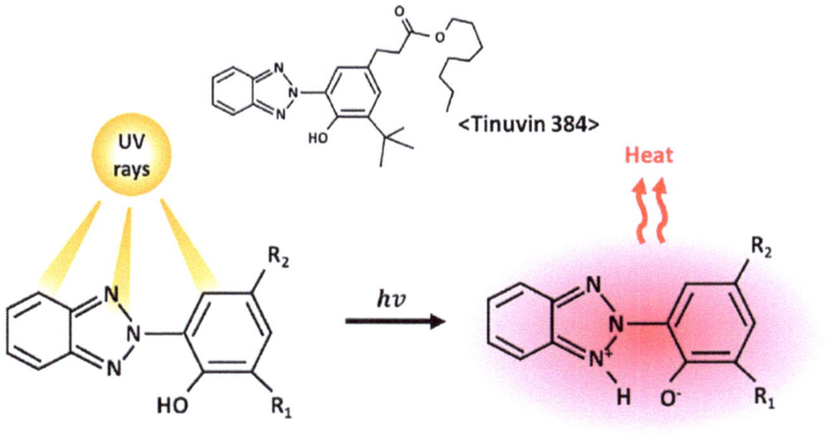

Fig. 4.9 Mechanism how UV absorber (Tinuvin UV absorber) dissipate heat that further causes photodegradation [60]

nanoparticles that added in PU coatings that act as reinforcement as well as UV shielding agent [62].

Studies reveal that silica nanoparticles introduced by proper methods reduce polymer degradation and help maintain gloss and mechanical integrity better than in situ polymerization methods [33]. Nanofillers like Nano-SiO₂, when combined with UV absorbers such as Tinuvin 292 or 384, enhance both mechanical durability and weathering resistance. Acrylic-PU coatings with Nano-SiO₂ plus stabilizers have demonstrated natural marine exposure lifespans exceeding 12 years [63]. CeO₂, TiO₂, and ZnO are all inorganic fillers absorbing UV light and act as fillers to reinforce the polymer matrix. CeO₂, in particular, is less toxic and effective at shielding. Experiments show significantly slowed photooxidation and reduced gloss loss under aging conditions when CeO₂ or TiO₂ is added in optimal amounts [64].

Polymer (Polyurethane) Structure Modification

This technique mitigates the effects of UV radiation and weathering by directly modifying the polymer's molecular structure. While traditional stabilizers like benzotriazoles or HALS (Hindered Amine Light Stabilizers) can reduce UV damage, structural modifications within the polymer backbone offer greater durability. These changes work by reducing UV-absorption sites or introducing UV-responsive bonds, making the polymer inherently more resistant to degradation [65]. These structural modifications done in multiple ways: incorporating UV-absorbing moieties in backbone, using dynamic bonds like disulfide units, polyol design (molecular weight and functionality), photo-crosslink able functional groups, and tannin-metal complex incorporation.

A study revealed that embedding benzophenone or benzotriazole units into the PU backbone (from di-functional UV-320 or methoxy benzophenone monomers) transforms incoming UV photons into thermal energy within the chain [57], thus, directly neutralizing UV energy before attack on urethane bonds. Such incorporation also improves yellowing resistance and tensile strength. Introducing disulfide bonds into hard segments creates self-reforming cycles under UV, UV cleaves –S–S– to sulfur radicals, which recombine dissipating energy and repairing damage simultaneously. This cycling reduces color change and mechanical deterioration [66]. Embedding coumarin units into the hard segments enables UV-triggered dimerization (at 354 nm) to crosslink chains, then reversal (cleavage at 254 nm) shown in Fig. 4.10. This reversible network can help maintain structure over weathering cycles [67]. Another study shows that in situ polymerization with tannic acid–Ti complexes (TATi) grafts bio-based phenolic UV absorbers into the waterborne PU chain. These strongly absorb UV and maintain surface integrity and adhesion after 1600 h artificial weathering.

4.7 Adhesion and Compatibility with Different Substrates

The adhesion and compatibility of polyurethane (PU) coatings with various substrates are crucial to their success as high-performance protective and decorative materials. Whether applied to metals, plastics, wood, glass, or concrete, the long-term durability and effectiveness of PU coatings largely depend on how well they bond with the underlying surface. While PU is widely valued for its mechanical strength and protective properties, its performance is directly influenced by proper adhesion and compatibility. Therefore, understanding substrate-specific surface preparation and formulation techniques is essential to ensure durable and reliable coating performance across different industrial applications. To under understand the adhesion and compatibility of PU coatings first, we have to understand the adhesion mechanism PU coatings.

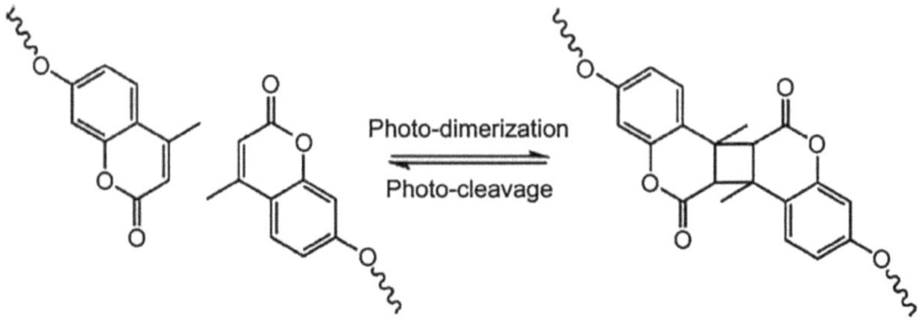

Fig. 4.10 Showing how photo-dimerization/photo-cleavage reactions occur in coumarin molecules and act as photo-crosslink able functional group [67]

4.7.1 Adhesion Mechanisms of PU Coatings

Adhesion between coating and substrate can occur through several mechanisms. First one is chemical bonding (Chemisorption); this involves formation of covalent or ionic bonds at the interface, often via reaction of isocyanate or hydroxyl/amine groups in PU with reactive groups on the substrate surface, especially significant for polar substrates like metals, glass, or oxidized plastics. This is quite effective way of adhesion and offers very strong adhesion [68]. Other mechanism of adhesion is secondary physical forces (physisorption); in this hydrogen bonding, dipole–dipole, and van der Waals forces are involving between substrate and coating especially on polar surfaces. Another adhesion mechanism is mechanical interlocking; this interlocking induced by physical anchoring where the coating infiltrates micro-roughness or pores of the substrate, leading to enhanced bond strength. Surface treatment (e.g., sandblasting for metals, sanding for wood) is applied to substrate because it is critical for this. Some other are diffusion and thermodynamic (wetting) that involves in the adhesion between coating and substrate [69]. Some adhesion mechanisms of PU coatings on different substrates are shown in Fig. 4.11.

4.7.2 Adhesion and Compatibility with Metal Substrate

Polyurethane (PU) coatings are widely employed across automotive, aerospace, marine, and infrastructure sectors due to their exceptional abrasion resistance, chemical stability, flexibility, and weatherability. However, the effectiveness of PU

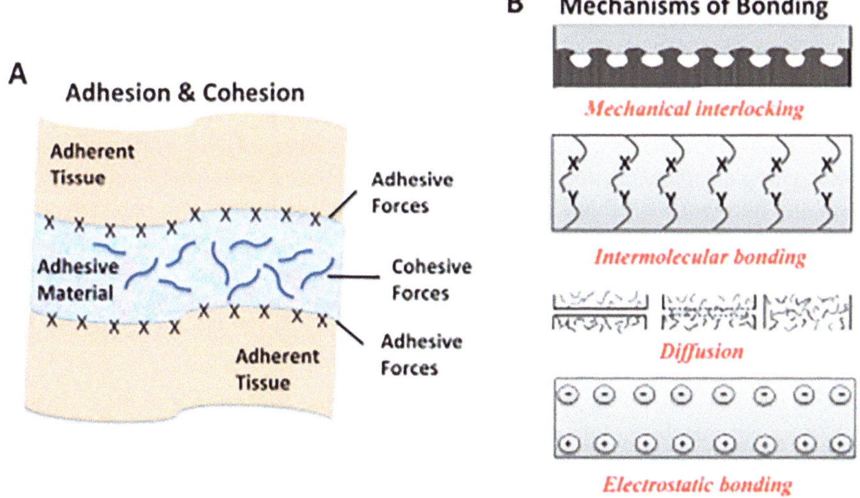

Fig. 4.11 Different adhesion mechanisms between coating layer and substrate [70]

coatings is not solely dependent on bulk material properties strong adhesion to the underlying metal substrate is critical to ensure long-term performance under environmental stressors like UV radiation, moisture, corrosive salts, and temperature cycling. The mechanisms of adhesion and compatibility are multifactorial, involving chemisorption, dispersive interactions, hydrogen bonding, and surface energy optimization all of which are now targets for advanced coating design.

PU prepolymers especially those derived from MDI (methylene diphenyl diisocyanate) can chemically bond to metal oxide surfaces (e.g., Fe_2O_3, Al_2O_3). These interactions form stable urethane–metal complexes and dramatically increase interfacial strength. A study by Nies et al. revealed that isocyanate moieties from MDI react directly with native aluminum and steel surfaces forming strong adhesive interphases, outperforming weaker polyether interactions [71]. Polyurethane segments (urethane, hydroxyl, or urea) interact with metal oxides and surface hydroxyls through secondary forces such as hydrogen bonding and dispersive interactions. These molecular-level forces contribute to adhesion especially where chemisorption is minimal. PU coatings with high –OH density exhibit stronger bonding with oxidized steel through hydrogen bridges. Organ silanes like γ-aminopropyltriethoxysilane (APS) or (3-glycidyloxypropyl) trimethoxysilane (GPTMS) act as chemical bridges, forming covalent Me–O–Si bonds on the metal side and reacting with urethane groups on the polymer side. This enhances adhesion and corrosion resistance. Silane-treated PU coatings retained >90% adhesion after 1000 h salt spray exposure, compared to 60% in untreated controls [8].

4.7.3   Adhesion and Compatibility with Plastics (Polymers)

Polyurethane (PU) coatings are increasingly used on polymeric substrates, such as PP, PE, PVC, PET, and TPU to impart desirable properties like abrasion resistance, UV protection, and chemical durability. However, adhesion to plastic surfaces is inherently challenging due to factors such as low surface energy, lack of polar chemical groups, and incompatibility of polymer polarity. Plastics have very low surface energy, making them hard to wet by PU coatings. Good adhesion requires the coating's surface tension to be lower than substrate energy, enabling close contact and interfacial bonding. Though chemical bonding is rare with non-polar plastics, PU's urethane groups can still form van der Waals and hydrogen bonds with treated or polarized surfaces, contributing to adhesion [72]. Overcoming these issues requires strategic surface treatments, formulation adjustments, and use of adhesion promoters or primers to enable reliable bonding and long-term coating performance.

PU coatings can strongly adhere to many polymer substrates if surface chemistry/energy is matched, often via surface modification or the introduction of compatibilizer molecules. For other plastics, optimize mechanical adhesion through surface roughening, increase chemical affinity by using compatible monomers or additives, and tailor PU formulation to balance flexibility and interfacial bonding [73].

Treatments like corona, plasma, flame, or chemical etching oxidize or functionalize polymer surfaces, increasing polarity and enabling improved wetting and bonding. Incorporation of nanoparticles, silanes, or acrylic moieties into the PU or substrate surface increases the interfacial area and chemical bridging, further improving adhesion and extending lifetime under demanding conditions [74, 75].

SFG spectroscopy and adhesion testing showed molecular-level interface ordering and stronger adhesion after plasma activation and thermal annealing, enhancing cohesive strength. PU coatings failed on untreated PE surfaces, while corona treatment significantly improved adhesion and allowed successful antibacterial PU films formulation on PE and TPU substrates. Use of primer also can increase the adhesion of PU coating and plastic substrate because primers contain organ silanes or reactive coupling agents create polar, reactive surface sites. When PU coatings are applied over these, adhesion dramatically improves even on low-energy plastics like PE and PP [76]. Mushtaq et al. studied and except that a two-layer system epoxy primer beneath PU topcoat on glass-epoxy composites delivered strong bonding evaluated via cross-hatch, 180° peel, and pull-off adhesion tests. Adhesion was superior when primer and substrate matched chemically and roughness optimized [77]. Wet PU adhesive layers (thin, nanometric) on glass-ITO electrodes allowed strong adhesion of PEDOT: PSS conducting polymer without compromising electrical/mechanical performance. Thin layers (~6–60 nm) enabled interpenetration, preserving low interface resistance and strong bonding [78].

4.7.4 Adhesion and Compatibility with Wood and Porous Substrates

Wood and other porous substrates such as bamboo, MDF, or natural fiber composites present unique challenges for PU coatings. While wood's porosity aids coating penetration and mechanical interlocking, it also introduces variability in surface chemistry, moisture content, and wettability. Ensuring durable adhesion of PU coatings to these substrates demands careful attention to surface preparation, penetration dynamics, polymer structure, and adhesion promoters [79]. The hydrophilic nature of wood also facilitates strong mechanical interlocking and hydrogen bonding with PU. However, extractives or surface contaminants may weaken the bond. Waterborne PU is especially successful for wood due to its ability to wet and penetrate the uneven substrate, improving adhesion and compatibility across grain, plywood, or MDF [80]. Adhesion peel rate of PU films on wood decreased to ~0.7% when using double-application sealing primer with primer and topcoat. The sealing primer penetrated pores, enhanced micro-groove formation, and improved physical interlocking and wetting [81].

On bamboo-based substrates, PU coatings showed contact angles around 30–31° on some surfaces and poor spread, resulting in adhesion classification 2 (5–15% flake area) on cross-cut tests. This shows wettability must be improved via substrate modification [82]. PU layer thickness can affect interfacial resistance and adhesion strength, particularly in sensitive conductive applications. Nano-thin coatings (6–60 nm) promote good adhesion without sacrificing function [78]. Higher crosslinking typically increases film hardness and mechanical strength, but the presence of adhesion promoters (such as silanes) can prevent loss of flexibility or bond strength [8].

4.7.5   Strategies to Improve Adhesion and Compatibility of PU Coating

Adhesive performance of PU coatings differs substantially across substrates like metals, plastics, wood, and composites. To achieve durable adhesion and compatibility, tailored approaches are required combining substrate preparation, coating chemistry, formulation additives, and promoters. The following study integrates mechanism-level insights, design principles, formulation, and surface engineering for improved PU adhesion on varied materials.

4.7.5.1   Surface Preparation and Activation

Surface activation transforms a substrate's outermost layer to improve wettability, adhesion, and chemical compatibility with PU coatings without affecting bulk properties. Typical methods include plasma, corona, or flame treatments, and occasionally chemical promoters. Mechanical roughening (sanding, bristle-blasting, micro-arc oxidation systems) increases surface area and promotes mechanical interlocking, especially on metals and composite substrates. For example, micro-arc oxidation (MAO) on aluminum alloy yields a porous ceramic sublayer that nearly doubles PU adhesion when further treated with silane coupling agents [83]. Research demonstrates that properly cleaned and roughened steel or aluminum substrates exhibit markedly better PU coating adhesion. Therefore, removal of oils, greases, and oxidation via solvent wiping or alkaline cleaning is mandatory before coating to ensure a contaminant-free interface [84]. Some advanced surface treatments are applied to substrates to achieve exceptional adhesion in high-end applications. Plasma, corona, and UV/ozone treatments are examples of such techniques.

Atmospheric plasma and corona treatments introduce polar functional groups (e.g., –OH, –COOH) onto non-reactive surfaces (plastics like PP and PE), dramatically enhancing surface energy and wettability by reducing its contact angle showing Fig. 4.12 [85]. Plasma treatments also disrupt organic contaminants and promote chemical bonding between coating and substrate by generating highly reactive radicals and functional groups. Seul et al. study the effect of plasma treatment on polyurethane-coated leather, they revealed that longer treatment time gives better adhesion and optical observation indicates surface change and wettability increase of the substrate after plasma treatment [86]. Particularly on polymers, UV/ozone

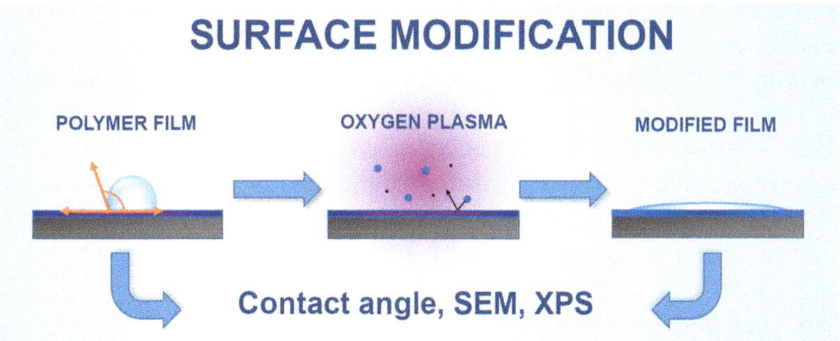

Fig. 4.12 Illustrating the plasma effect on substrate surface. Plasma treatment decreases the contact angle that causes increase in wettability of the surface, results are confirmed by SEM and XPS [88]

activation increases the presence of oxygen-containing functional groups, which foster strong chemical and physical attraction for PU coatings [87].

A helium DBD-plasma treatment applied to PU film on glass reduced water contact angle from 90° to 79°, increased critical surface tension by 2 dynes/cm, and led to a tenfold increase in endothelial-cell adhesion. Peel strength and wettability improvement were directly linked to surface oxidation and polarity increase. Some limitations of plasma treatment also arise when applying oxygen plasma to PU coil coatings, pores and micro-defects may form, degrading barrier properties. Comparatively, argon plasma activation is less destructive. Fluorine-containing precursors mitigate damage and maintain barrier function [89].

4.7.5.2 Application of Primer Layers and Tie Coats

Primer layers and tie coats are applied to substrates to significantly improve the adhesion of polyurethane (PU) coatings by creating strong chemical and physical bonds between the substrate and the coating. A primer layer is a surface-modifying coating applied directly to the substrate. It contains functional groups that chemically react with both the substrate surface and the subsequent PU coating, forming robust covalent or polar bonds. Essentially, primers act as a molecular bridge that bonds well to both the substrate and the coating. Tie coats (or tie layers), on the other hand, are layers often polymeric resins with functionalized chemical groups applied between the substrate and the PU coating [90]. Tie coats are typically designed to bond chemically to the substrate on one side and physically or chemically to the PU coating on the other. Together, primers and tie coats complement each other, the primer conditions the substrate surface to enhance chemical bonding, while the tie coat provides an adhesive interlayer that improves mechanical and chemical compatibility with the PU coating. This combination results in extraordinary adhesion strength, durability, and resistance to environmental factors in high-end applications of PU coatings [91].

Chlorinated polyolefins (CPOs) are effective primers on low-energy plastics like PP and PE, enabling adhesion at thin film thickness (<5 μm) and improving compatibility with PU formulations. For metals, zinc-, phosphate-, or silane-based primers create interfacial layers that bond strongly to both the substrate and the PU, improving both adhesion and corrosion resistance. While tie coat formulations allow for reliable application of new PU coatings onto aged or difficult surfaces without mechanical abrasion, as shown in aerospace refinishing [92].

4.7.5.3 Use of Adhesion Promoters

Adhesion promoters in polyurethane (PU) coatings are specialized chemicals added to improve the bond between the coating and the substrate. They function primarily by creating a chemical "bridge" or interface that connects the substrate surface and the PU coating at a molecular level. This is especially useful for substrates with low surface energy or low polarity, which naturally have poor adhesion to coatings.

Silane coupling agent is one of the famous adhesion promoters in the coating industry. Silane coupling agents (like amino-silanes, glycidyloxypropyl- and trimethoxysilyl derivatives) form molecular bridges between inorganic surfaces (metals, glass, oxides) and the organic PU matrix, providing robust chemical and sometimes even covalent bonding [93]. A study predicts that formulations with 43 wt.% TMSCH exhibit enhanced scratch resistance and surface wettability. Recent patents and research reports describe custom acrylate-styrene and organosilicon-based adhesion promoters for PU, especially for boosting bonding to metals and thermoplastics, with notable improvements in peel and cross-cut adhesion tests. Inclusion of multifunctional resins or compatible reactive groups in the PU structure itself (e.g., isocyanate-terminated chains with pendant silane, acrylate, or epoxy groups) allows chemical grafting directly at the interface. Blending with acrylic, phenolic, or chlorinated polymers can optimize both cost and substrate compatibility, especially for contact or heat-activated PU adhesives across mixed substrates [94].

References

1. Wu S et al (2025) A comprehensive review of polyurethane: properties, applications and future perspectives. Polymer 327:128361. https://doi.org/10.1016/j.polymer.2025.128361
2. Fang Z, Huang L, Fu J (2022) Research status of graphene polyurethane composite coating. Coatings 12(2):264. https://doi.org/10.3390/coatings12020264
3. The influence of soft segment structure on the properties of polyurethanes. https://www.mdpi.com/2073-4360/15/18/3755. Accessed 17 Aug 2025
4. Yang SJ, Rosenbloom SI, Fors BP, Silberstein MN (2024) Elucidating the impact of microstructure on mechanical properties of phase-segregated polyurea: finite element modeling of molecular dynamics derived microstructures. Mech Mater 188:104863. https://doi.org/10.1016/j.mechmat.2023.104863

5. Patti A, Costa F, Perrotti M, Barbarino D, Acierno D (2021) Polyurethane impregnation for improving the mechanical and the water resistance of polypropylene-based textiles. Materials 14(8):1951. https://doi.org/10.3390/ma14081951

6. Samad A, Siew WH, Given M, Liggat J, Timoshkin I (2024) Effects of polyurethane hardness on the propagation of acoustic signals from partial discharge. High Volt 9(5):1125–1135. https://doi.org/10.1049/hve2.12477

7. Improving water resistance and mechanical properties of crosslinked waterborne polyurethane using glycidyl carbamate. https://www.mdpi.com/2073-4360/16/19/2794. Accessed 17 Aug 2025

8. Kant İE, Tarkuç S, Köken N, Yeşilçubuk SA, Kızılcan N (2024) Enhancement of the surface and mechanical properties of polyurethane coating through changing the additive ratio of Silane. Adv Mater Interfaces 11(11):2300944. https://doi.org/10.1002/admi.202300944

9. Improvement of mechanical properties and solvent resistance of polyurethane coating by chemical grafting of graphene oxide. https://www.mdpi.com/2073-4360/15/4/882. Accessed 17 Aug 2025

10. Navidfar A (2021) Fabrication and characterization of hybrid nanofiller reinforced polyurethane nanocomposites. http://hdl.handle.net/11527/20144. Accessed 17 Aug 2025

11. Youngblood JP, Chowdhury RA, Azrak SMEA (2023) Cellulose nanomaterial (CN) based waterborne polyurethane coating. US11827795B2. https://patents.google.com/patent/US11827795/en. Accessed 17 Aug 2025

12. Rajabimashhadi Z, Naghizadeh R, Zolriasatein A, Bagheri S, Mele C, Esposito Corcione C (2023) Hydrophobic, mechanical, and physical properties of polyurethane nanocomposite: synergistic impact of Mg(OH)2 and SiO2. Polymers 15(8):1916. https://doi.org/10.3390/polym15081916

13. Wang R et al (2023) Advanced progress on the significant influences of multi-dimensional nanofillers on the tribological performance of coatings. RSC Adv 13(29):19981–20022. https://doi.org/10.1039/D3RA01550E

14. Li X, Nikiforow I, Pohl K, Adams J, Johannsmann D (2013) Polyurethane coatings reinforced by halloysite nanotubes. Coatings 3(1):16–25. https://doi.org/10.3390/coatings3010016

15. Bekbayeva L et al (2025) Optimizing anticorrosion coating performance: synthesis of polyurethane/epoxy hybrids. Polymers 17(11):1516. https://doi.org/10.3390/polym17111516

16. Chen S, Tian Y, Chen L, Hu T (2006) Epoxy resin/polyurethane hybrid networks synthesized by frontal polymerization. Chem Mater 18(8):2159–2163. https://doi.org/10.1021/cm052391s

17. Wang T et al (2021) Interlayer molecular migration and reaction in an epoxy-polyurethane coating system: implications for the system hardness. Prog Org Coat 151:106083. https://doi.org/10.1016/j.porgcoat.2020.106083

18. Optimizing anticorrosion coating performance: synthesis of polyurethane/epoxy hybrids. https://www.mdpi.com/2073-4360/17/11/1516. Accessed 21 Aug 2025

19. Mechanism confirmation of organofunctional silanes modified sodium silicate/polyurethane composites for remarkably enhanced mechanical properties. ResearchGate. https://www.researchgate.net/publication/351247704_Mechanism_confirmation_of_organofunctional_silanes_modified_sodium_silicatepolyurethane_composites_for_remarkably_enhanced_mechanical_properties. Accessed 18 Aug 2025

20. Liu M et al (2025) Synthesis and characterization of silane-coupled sodium silicate composite coatings for enhanced anticorrosive performance. Coatings 15(4):428. https://doi.org/10.3390/coatings15040428

21. Lyu J et al (2019) In situ incorporation of diamino silane group into waterborne polyurethane for enhancing surface hydrophobicity of coating. Molecules 24(9):1667. https://doi.org/10.3390/molecules24091667

22. Zafar S, Kahraman R, Shakoor RA (2024) Recent developments and future prospective of polyurethane coatings for corrosion protection—a focused review. Eur Polym J 220:113421. https://doi.org/10.1016/j.eurpolymj.2024.113421

23. Das S et al (2023) SERS nanowire chip and machine learning-enabled classification of wild-type and antibiotic-resistant bacteria at species and strain levels. ACS Appl Mater Interfaces 15(20):24047–24058. https://doi.org/10.1021/acsami.3c00612

24. Effect of various reactive diluents on the mechanical properties of the acrylate-based polymers produced by DLP/LCD-type 3D printing. ResearchGate. https://www.researchgate.net/publication/381838806_Effect_of_various_reactive_diluents_on_the_mechanical_properties_of_the_acrylate-based_polymers_produced_by_DLPLCD-type_3D_printing. Accessed 18 Aug 2025

25. Ahmed MA, Shohide MA, El-Saeed AM, Naguib HM (2024) Coating effect of polyurethane-layered double hydroxide nanocomposite on steel. Discov Mater 4(1):42. https://doi.org/10.1007/s43939-024-00109-2

26. Li O et al (2023) Flexible polyurethane foams reinforced with graphene and boron nitride nanofillers. Polym Compos 44(3):1494–1511. https://doi.org/10.1002/pc.27183

27. Qin Y et al (2024) Osthole-infused polyurethane flexible coatings for enhanced underwater drag reduction and robust anti-biofouling. Prog Org Coat 188:108213. https://doi.org/10.1016/j.porgcoat.2024.108213

28. Xie C, Shi Y, Wu P, Sun B, Yue Y (2024) Influence of dilution on the mechanical properties and microstructure of polyurethane-cement based composite surface coating. Polymers 16(1):146. https://doi.org/10.3390/polym16010146

29. Guo H, Yin M, Lv X, Chen Y, Sun M (2023) High-strength and high-toughness spray-type poly-urea coating for impact protection of aluminum sheet. J Coat Technol Res 20(6):2053–2068. https://doi.org/10.1007/s11998-023-00801-7

30. Sheikhi MR, Hasanzadeh M, Gürgen S, Li J (2024) Enhanced anti-impact resistance of polyurethane foam composites with multi-phase shear thickening fluids containing various carbon nanofillers. Mater Today Commun 38:107991. https://doi.org/10.1016/j.mtcomm.2023.107991

31. Ghosh A et al (2023) Polyurethane chemistry for the agricultural applications—recent advancement and future prospects. In: Polyurethanes: preparation, properties, and applications, Emerging applications, vol 3. American Chemical Society, pp 1–36. https://doi.org/10.1021/bk-2023-1454.ch001

32. Wang X, Hu J, Li Y, Zhang J, Ding Y (2015) The surface properties and corrosion resistance of fluorinated polyurethane coatings. J Fluor Chem 176:14–19. https://doi.org/10.1016/j.jfluchem.2015.04.002

33. Weathering resistance of waterborne polyurethane coatings reinforced with silica from rice husk ash. ResearchGate. https://www.researchgate.net/publication/337477173_Weathering_Resistance_of_Waterborne_Polyurethane_Coatings_Reinforced_with_Silica_from_Rice_Husk_Ash. Accessed 18 Aug 2025

34. Ravindran Marwan Shalash EN, Al Azzam KM, Sagatbekovna MZ, Kairatovna BA, Konstantinovich AT, Adlikhanovna NA, Ermekovna KA, Balasubramani (2024) Synthesis, characterization, and application of polyurethane-acrylic hybrids as anticorrosion coatings. Int J Technol. https://ijtech.eng.ui.ac.id/article/view/7044. Accessed 17 Aug 2025

35. Rayung M, Ghani NA, Hasanudin N (2024) A review on vegetable oil-based non isocyanate polyurethane: towards a greener and sustainable production route. RSC Adv 14(13):9273–9299. https://doi.org/10.1039/D3RA08684D

36. Chhipa SM, Sharma S, Kumar Bagha A (2024) Recent development in polymer coating to prevent corrosion in metals: a review. Mater Today Proc. https://doi.org/10.1016/j.matpr.2024.09.001

37. Combined concurrent physical and chemical model for accelerated weathering damages of polyurethane-based coatings. ResearchGate. https://www.researchgate.net/publication/385317225_Combined_concurrent_Physical_and_Chemical_model_for_accelerated_weathering_damages_of_polyurethane-based_coatings. Accessed 18 Aug 2025

38. Jia X, Song X, Fan J, Sun W, Liu C (2022) Research Progress in characterization methods of anti-corrosion and Wear-resistant polyurethane coatings. MATEC Web Conf 358:01045. https://doi.org/10.1051/matecconf/202235801045

39. Yeligbayeva G et al (2024) Polyurethane as a versatile polymer for coating and anti-corrosion applications: a review. Kompleks. Ispol'z Miner Syra' = Complex Use Miner. Resour 331(4):21–41. https://doi.org/10.31643/2024/6445.36

40. Li X et al (2025) A biomass green synthesis process: aqueous jatropha oil-based polyurethane coating with high transparency, hydrophobicity and corrosion resistance. Prog Org Coat 206:109303. https://doi.org/10.1016/j.porgcoat.2025.109303

41. Liu J, Recupido F, Lama GC, Oliviero M, Verdolotti L, Lavorgna M (2023) Recent advances concerning polyurethane in leather applications: an overview of conventional and greener solutions. Collagen Leather 5(1):8. https://doi.org/10.1186/s42825-023-00116-8

42. Lopes RVV et al (2025) Accelerated weathering tests of linseed and passion fruit oil-based polyurethanes. Coatings 15(5):550. https://doi.org/10.3390/coatings15050550

43. Huang G et al (2023) Colorless, transparent, and high-performance polyurethane with intrinsic ultraviolet resistance and its anti-UV mechanism. ACS Appl Mater Interfaces 15(14):18300–18310. https://doi.org/10.1021/acsami.2c23317

44. Majidi R, Keramatinia M, Ramezanzadeh B, Ramezanzadeh M (2023) Weathering resistance (UV-shielding) improvement of a polyurethane automotive clear-coating applying metal-organic framework (MOF) modified GO nano-flakes (GO-ZIF-7). Polym Degrad Stab 207:110211. https://doi.org/10.1016/j.polymdegradstab.2022.110211

45. Liu F, Hao Y, Wang Z, Shi H, Han E, Ke W (2010) Flaking and degradation of polyurethane coatings after 2 years of outdoor exposure in Lhasa. Chin Sci Bull 55(7):650–655. https://doi.org/10.1007/s11434-009-0269-1

46. Albergamo V, Wohlleben W, Plata DL (2023) Photochemical weathering of polyurethane microplastics produced complex and dynamic mixtures of dissolved organic chemicals. Environ Sci Process Impacts 25(3):432–444. https://doi.org/10.1039/D2EM00415A

47. Feldman D (2002) Polymer weathering: photo-oxidation. J Polym Environ 10(4):163–173. https://doi.org/10.1023/A:1021148205366

48. Kotnarowska D (2018) Influence of ageing with UV radiation on physicochemical properties of acrylic-polyurethane coatings. J Surf Eng Mater Adv Technol 8(4):95–109. https://doi.org/10.4236/jsemat.2018.84009

49. Mayer-Trzaskowska P, Robakowska M, Gierz Ł, Pach J, Mazur E (2024) Observation of the effect of aging on the structural changes of polyurethane/polyurea coatings. Polymers 16(1):23. https://doi.org/10.3390/polym16010023

50. Acierno D, Graziosi L, Patti A (2023) Puncture resistance and UV aging of nanoparticle-loaded waterborne polyurethane-coated polyester textiles. Materials 16(21):6844. https://doi.org/10.3390/ma16216844

51. Rosu D, Rosu L, Cascaval CN (2009) IR-change and yellowing of polyurethane as a result of UV irradiation. Polym Degrad Stab 94(4):591–596. https://doi.org/10.1016/j.polymdegradstab.2009.01.013

52. Theiler G, Wachtendorf V, Elert A, Weidner S (2018) Effects of UV radiation on the friction behavior of thermoplastic polyurethanes. Polym Test 70:467–473. https://doi.org/10.1016/j.polymertesting.2018.08.006

53. Liu J et al (2019) Degradation behavior and mechanism of polyurethane coating for aerospace application under atmospheric conditions in South China Sea. Prog Org Coat 136:105310. https://doi.org/10.1016/j.porgcoat.2019.105310

54. NATURAL SUBSTANCES AS PROTECTIVE AGENTS AGAINST PHOTODEGRADATION. ResearchGate. https://www.researchgate.net/publication/385464213_NATURAL_SUBSTANCES_AS_PROTECTIVE_AGENTS_AGAINST_PHOTODEGRADATION. Accessed 18 Aug 2025

55. UV Absorber UV-400 for long-term color stability and mechanical integrity in polymers_ Toluene diisocyanate manufacturer. https://www.allhdi.com/archives/72619. Accessed 18 Aug 2025

56. Jones BJP, VanGemert JK, Conrad JM, Pla-Dalmau A (2013) Photodegradation mechanisms of tetraphenyl butadiene coatings for liquid argon detectors. J Instrum 8(01):P01013. https://doi.org/10.1088/1748-0221/8/01/P01013

57. Oprea S, Potolinca VO (2022) UV protection by the inclusion of the methoxybenzophenone moieties into the backbone chain of the polyurethane structure. J Polym Res 29(9):369. https://doi.org/10.1007/s10965-022-03230-z

58. Li J, Sandlin S, Moncur M (2019) Durable, electrically conductive transparent polyurethane compositions and methods of applying same. US20190153259A1. https://patents.google.com/patent/US20190153259A1/en. Accessed 18 Aug 2025

59. Qiao R-M, Zhao C-P, Liu J-L, Zhang M-L, He W-Q (2022) Synthesis of novel ultraviolet absorbers and preparation and field application of anti-ultraviolet aging PBAT/UVA films. Polymers 14(7):1434. https://doi.org/10.3390/polym14071434

60. Yang D, Ramu AG, Choi D (2022) Synthesis of transparent ZnO–TiO2 and its nanocomposites for ultraviolet protection of a polyethylene terephthalate (PET) film. Catalysts 12(12):1590. https://doi.org/10.3390/catal12121590

61. Nanoparticles and their applications in ultraviolet protection: a review. ResearchGate. https://www.researchgate.net/publication/225084634_Nanoparticles_and_their_applications_in_ultraviolet_protection_A_Review. Accessed 18 Aug 2025

62. Xiao J et al (2023) Polymer/TiO2 hybrid nanoparticles with highly effective UV-screening but eliminated photocatalytic activity. Macromolecules 46(2):375–383. https://doi.org/10.1021/ma3022019

63. The synergistic effects of Sio2 nanoparticles and organic photostabilizers for enhanced weathering resistance of acrylic polyurethane coating. https://www.mdpi.com/2504-477X/4/1/23. Accessed 18 Aug 2025

64. Multi self-healable UV shielding polyurethane/ceo2 protective coating: the effect of low-molecular-weight polyols. https://www.mdpi.com/2073-4360/12/9/1947. Accessed 18 Aug 2025

65. Czachor-Jadacka D, Pilch-Pitera B, Kisiel M, Thomas J (2023) Polyurethane powder coatings with low curing temperature: research on the effect of chemical structure of crosslinking agent on the properties of coatings. Prog Org Coat 182:107662. https://doi.org/10.1016/j.porgcoat.2023.107662

66. Preparation of yellowing-resistant waterborne polyurethane modified with disulfide bonds. https://www.mdpi.com/1420-3049/29/9/2099. Accessed 18 Aug 2025

67. Germán-Ayuso L et al (2023) Improving the performance of biobased polyurethane dispersion by the incorporation of photo-crosslinkable coumarin. J Coat Technol Res 20:1677–1690. https://doi.org/10.1007/s11998-023-00772-9

68. Wu Y et al (2023) Molecular behavior of 1K polyurethane adhesive at buried interfaces: plasma treatment, annealing, and adhesion. Langmuir 39(9):3273–3285. https://doi.org/10.1021/acs.langmuir.2c03084

69. Study on the effect of surface energy of polypropylene/polyamide12 polymer hybrid matrix reinforced with virgin and recycled carbon fiber. ResearchGate. https://www.researchgate.net/publication/323199567_Study_on_the_Effect_of_Surface_Energy_of_PolypropylenePolyamide12_polymer_Hybrid_Matrix_Reinforced_with_Virgin_and_Recycled_Carbon_Fiber. Accessed 18 Aug 2025

70. Bal-Ozturk A et al (2021) Tissue adhesives: from research to clinical translation. Nano Today 36:101049. https://doi.org/10.1016/j.nantod.2020.101049

71. Adhesive interactions of polyurethane monomers with native metal surfaces. ResearchGate. https://www.researchgate.net/publication/235920518_Adhesive_Interactions_of_Polyurethane_Monomers_with_Native_Metal_Surfaces. Accessed 18 Aug 2025

72. Awaja F, Gilbert M, Kelly G, Fox B, Pigram PJ (2009) Adhesion of polymers. Prog Polym Sci 34(9):948–968. https://doi.org/10.1016/j.progpolymsci.2009.04.007

73. Kaur R, Verma SK, Mehta R (2025) Tailoring the properties of polyurethane composites: a comprehensive review. Polym-Plast Technol Mater 64(13):2004–2018. https://doi.org/10.1080/25740881.2025.2493869
74. Khdheer MF et al (2025) Preparation and characterization of polyurethane composite coatings reinforced with carbon waste. Polym Adv Technol 36(7):e70240. https://doi.org/10.1002/pat.70240
75. Review on the use of nanofillers in polyurethane coating systems for different coating applications. ResearchGate. https://www.researchgate.net/publication/338797107_Review_on_the_Use_of_Nanofillers_in_Polyurethane_Coating_Systems_for_Different_Coating_Applications. Accessed 18 Aug 2025
76. Morgese G, Siegmann K, Winkler M (2024) Specific, nondestructive, and durable adhesion primer for polyolefins. J Coat Technol Res 21(6):1921–1930. https://doi.org/10.1007/s11998-024-00938-z
77. Mushtaq M, Adusumalli RB, Suresh K, Anna Abraham A (2023) Adhesion and tribological characteristics of modified polyurethane coating on composite substrate. Surf Eng 39(7–12):836–851. https://doi.org/10.1080/02670844.2023.2264528
78. Inoue A, Yuk H, Lu B, Zhao X (2020) Strong adhesion of wet conducting polymers on diverse substrates. Sci Adv 6(12):eaay5394. https://doi.org/10.1126/sciadv.aay5394
79. Kúdela J, Liptáková E (2006) Adhesion of coating materials to wood. J Adhes Sci Technol 20(8):875–895. https://doi.org/10.1163/156856106777638725
80. Waterborne polyurethane with excellent UV resistance and strong adhesion modified with tannic acid metal complexes. https://www.google.com/search?q=Waterborne+polyurethane+with+excellent+UV+resistance+and+strong+adhesion+modified+with+tannic+acid+metal+complexes&rlz=1C1FHFK_en-GBPK1170PK1170&oq=Waterborne+polyurethane+with+excellent+UV+resistance+and+strong+adhesion+modified+with+tannic+acid+metal+complexes&gs_lcrp=EgZjaHJvbWUqBggAEEUYOzIGCAAQRRg7MgYIARBFGD0yBggCEEUYPTIGCAMQRRg90gEKMTQ0NTY1ajBqN6gCALACAA&sourceid=chrome&ie=UTF-8. Accessed 18 Aug 2025
81. Effect of coating process on properties of two-component waterborne polyurethane coatings for wood. ResearchGate. https://www.researchgate.net/publication/365880193_Effect_of_Coating_Process_on_Properties_of_Two-Component_Waterborne_Polyurethane_Coatings_for_Wood. Accessed 18 Aug 2025
82. Xu J, Liu R, Wu H, Qiu H, Yu Y, Long L (2019) Coating performance of water-based polyurethane-acrylate coating on bamboo/bamboo scrimber substrates. Adv Polym Technol 2019(1):4264701. https://doi.org/10.1155/2019/4264701
83. Wang R, Xu H, Yao Z, Li C, Jiang Z (2020) Adhesion and corrosion resistance of micro-arc oxidation/polyurethane composite coating on aluminum alloy surface. Appl Sci 10(19):6779. https://doi.org/10.3390/app10196779
84. Mayer P, Lubecki M, Stosiak M, Robakowska M (2022) Effects of surface preparation on the adhesion of UV-aged polyurethane coatings. Int J Adhes Adhes 117:103183. https://doi.org/10.1016/j.ijadhadh.2022.103183
85. Duzcukoglu H, Kaybal HB, Asmatulu R (2025) Surface treatment strategies for improving adhesion of PEDOT:PSS coatings on aerospace fiber composites. Polym Bull 82(11):5543–5570. https://doi.org/10.1007/s00289-025-05752-0
86. Seul SD, Lim JM, Ha SH, Kim YH (2005) Adhesion enhancement of polyurethane coated leather and polyurethane foam with plasma treatment. Korean J Chem Eng 22(5):745–749. https://doi.org/10.1007/BF02705793
87. Hamdi M, Saleh MN, Poulis JA (2020) Improving the adhesion strength of polymers: effect of surface treatments. J Adhes Sci Technol 34(17):1853–1870. https://doi.org/10.1080/01694243.2020.1732750
88. Mrsic I et al (2021) Oxygen plasma surface treatment of polymer films—pellethane 55DE and EPR-g-VTMS. Appl Surf Sci 536:147782. https://doi.org/10.1016/j.apsusc.2020.147782

89. Sundelacruz S, Kaplan DL (2009) Stem cell- and scaffold-based tissue engineering approaches to osteochondral regenerative medicine. Semin Cell Dev Biol 20(6):646–655. https://doi.org/10.1016/j.semcdb.2009.03.017
90. Regnier B (2023) Polyurethane-based primer for enhancing adhesion of liquid toner. WO2013126869A1. https://patents.google.com/patent/WO2013126869A1/en. Accessed 18 Aug 2025
91. Allen R. Using primers in combination with adhesive tie-layer resins or their blends to make structures with unique performance
92. Aklian J, Tang G, Abrami S (2024) Polyurethane coatings with improved interlayer adhesion. CA2698777C. https://patents.google.com/patent/CA2698777C/en. Accessed 18 Aug 2025
93. Sivakumar P, Du SM, Selter M, Daye J, Cho J (2021) Improved adhesion of polyurethane-based nanocomposite coatings to tin surface through silane coupling agents. Int J Adhes Adhes 110:102948. https://doi.org/10.1016/j.ijadhadh.2021.102948
94. Three-dimensional printing, wearables, medical textiles, adhesives, and coatings. ResearchGate. https://www.researchgate.net/publication/357845558_Three-Dimensional_Printing_Wearables_Medical_Textiles_Adhesives_and_Coatings. Accessed 18 Aug 2025

Chapter 5
Applications for Polyurethane Coatings

5.1 Industrial Applications (Automotive, Aerospace, Construction, Etc.)

Polyurethane (PU) coatings are extensively applied across numerous industrial sectors due to their superior mechanical properties, chemical resistance, flexibility, and aesthetics. Their versatility has made them a preferred choice for protective and decorative coatings in automotive, aerospace, construction, marine, furniture, electronics, and other emerging industries shown in Fig. 5.1 [1]. Polyurethane's qualities can vary greatly, ranging from rock-hard, unyielding construction material to soft-touch coatings. The simplicity of customization and these mechanical, chemical, and biological qualities have generated a great deal of interest among the interested parties as well as the scientific community and industry as well. The material can be improved by adding various additives and nanomaterials, as well as by modifying the raw elements. Therefore, a polyurethane that is practically appropriate for every application can be produced by making the right changes to the basic material [2]. The following is the detail of the industrial applications of PU coatings, to discuss their functional roles, advantages, challenges, and innovations tailored for each sector.

5.2 PU Coatings in Automotive and Transportation Industry

PU coatings in automotive are used as primers, basecoats, clearcoats, and interior finishes. Their high gloss, chemical resistance, weather durability, and color retention make them ideal for exteriors, while low-VOC waterborne PU is preferred for interiors. PU coatings withstand washing, road salts, UV aging, and abrasions from flexing panels. Modern vehicles increasingly incorporate bio-based polyurethane

© The Author(s), under exclusive license to Springer Nature
Switzerland AG 2026

Z. Zubair et al., *Functional Polyurethane Coatings*, SpringerBriefs in Materials,
https://doi.org/10.1007/978-3-032-12730-3_5

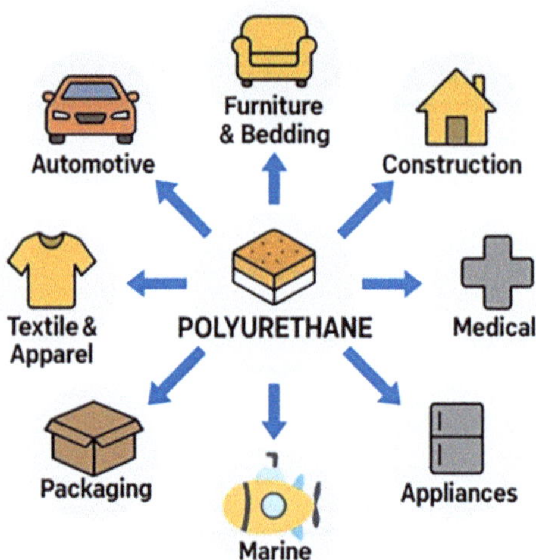

Fig. 5.1 Different application areas of PU coatings [1]

foams for interior trim, seating, cushioning, and thermal insulation. Ford and others adopt soybean-derived polyols to reduce environmental impact while maintaining performance [3]. Thinner PU coatings with high efficacy contribute to vehicle weight reduction, improving fuel economy and reducing emissions. Research shows that PU coatings formulated for electric vehicles (EVs) protect sensitive battery compartments and electronic parts from thermal and chemical stresses [4]. Advances in nano-reinforced as we discussed in mechanical properties section, PU coatings incorporating materials like graphene oxide and silica nanoparticles have improved abrasion resistance and toughness, extending wear life under harsh road conditions. Axalta and BASF have released PU coating solutions focusing on durability, scratch resistance, and prolonged gloss retention suitable for automotive applications. A detailed market report highlights continuous growth driven by these innovations [5].

Although polyurethane-based paints and coatings are not a novel technology, the advent of nanoparticles and nanofillers has lately created a new area for improving certain properties. Verma et al. investigated the interaction between the fillers and the hard segments. He demonstrated the favorable interaction between intercalated and exfoliated clay platelets and the rigid polyurethane resin segments [6]. Isocyanates are essential components in modern automotive clearcoats, offering key benefits such as weather resistance, scratch protection, and mechanical durability through their polyurethane structure. Their viscosity can be tailored to enable high solids formulations, helping reduce VOC emissions in line with environmental regulations. Due to their versatility and performance, isocyanates remain critical to the development of advanced, compliant automotive coatings [7]. The remarkable success of polyurethane (PU) coatings is largely due to their inherent chemical versatility, particularly the reactivity of isocyanate groups with various partners, enabling diverse crosslinking options. This results in robust, stable materials suitable for a wide range of applications and service temperatures (−60 to 250 °C) across solvent-borne, solvent-free, and waterborne systems. PU coatings can be precisely tailored to specific needs flexibility for plastics, cans, and coils, or durability and impact resistance for automotive clearcoats [8]. Three different aliphatic isocyanates are used as the primary building blocks for lightfast PU coatings; hexamethylene diisocyanate (HDI), isophorone diisocyanate (IPDI), and 4,4′-dicyclohexylmethane diisocyanate (H12MDI). Penta-methylene diisocyanate (PDI) was introduced in 2015. It is a bio-based alternative to HDI-derived coating materials, leading to similar or even improved PU coating properties in automotive application [9]. It uses in automotive because it imparts following characteristics on to substrate that play very crucial part in enhancing the life, aesthetic look, and performance of the automotive.

5.2.1   Scratch Resistance in Automotive

PU-based automotive clearcoats typically consist of acrylic-polyurethane or polyurethane hybrids, offer superior scratch and mar resistance compared to alternatives. A study found that PU clearcoats maintain better scratch recovery even after prolonged weathering, outperforming acrylic/melamine/silane systems under high-temperature and UV exposure [10]. Another study analyzed how polyrotaxane and silane affect scratch properties of clear coatings. They found that adding silane-functionalized molecular necklace crosslinkers enhances critical load resistance, likely due to improved crosslinking and dynamic motion from polyrotaxane. Environmental effects on scratch durability are also examined [11]. Figure 5.2 representing that how the coating on a substrate increases the scratch resistance.

Research has extensively explored scratch and mar resistance using advanced nano-/micro-scratch tests and real-world simulations like car washes. Findings show that increasing rubber-like elasticity and reducing interchain interactions enhances scratch resistance. Coatings with silane-modified blocked isocyanate crosslinkers demonstrate superior scratch and recovery performance in both lab

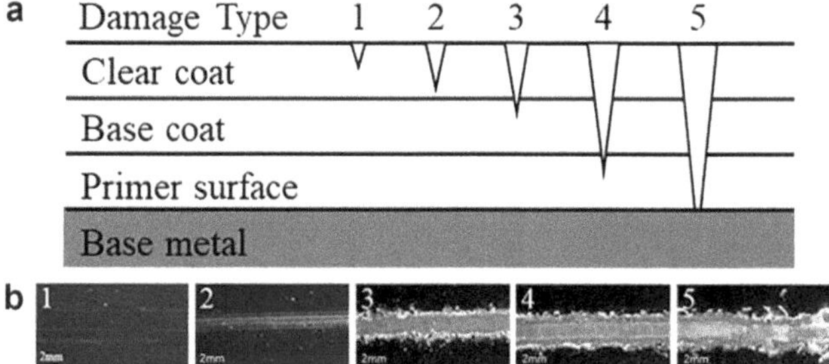

Fig. 5.2 Illustration of progressive coating damage types and corresponding microscopic features: (**a**) schematic diagram showing the depth of damage penetration from the clear coat to the base metal and (**b**) microscopic images representing damage severity level [11]

tests and real-world conditions. Nano-scratch techniques effectively replicate wear scenarios, helping correlate crosslink density with mechanical durability [12]. Industry reports further support these findings, with modified PU clearcoats maintaining 80–90% of original gloss even after intense simulated carwash and weathering exposure, due to higher crosslink density and SiO_2-like hard domains in the coating matrix [13]. Another approach explored waterborne UV-curable PU acrylate clearcoats, using nano ZnO as photo-initiators offering rapid curing, reduced VOCs, and potential use in sustainable automotive applications [14].

5.2.2 *Corrosion Resistance in Automotive*

Polyurethane (PU) coatings are extensively used in the automotive industry to enhance the appearance and provide corrosion resistance to vehicles. Additionally, PU coatings enhance corrosion resistance by limiting electrolyte access to metals. These coatings are therefore essential for protecting automotive components from environmental damage and extending their lifespan.

PU coatings embedded with microcapsules containing linseed oil loaded with sodium dodecyl sulfate (SDS) act as corrosion inhibitors that are released upon damage. The formulation with 4 wt.% microcapsules achieved over 90% inhibition efficiency, demonstrated self-healing in scratched coatings, and significantly improved water uptake resistance versus capsules with only linseed oil [15]. These coatings incorporate magnetic microcapsules structured in a biomimetic interface to promote uniform dispersion and targeted self-healing. They prolonged the corrosion onset by 816 h and boosted antifouling performance approximately 30-fold compared to controls [16]. A study by Wen et al. demonstrated that embedding low levels (~0.3 wt.%) of functionalized graphene oxide (IP-GO) into waterborne PU substantially increased electrochemical impedance and delayed electrolyte penetration, enhancing corrosion resistance as shown in Fig. 5.3 even after 50 days in a 3.5% NaCl test [17]. Another study revealed that GO-functionalized PU coatings,

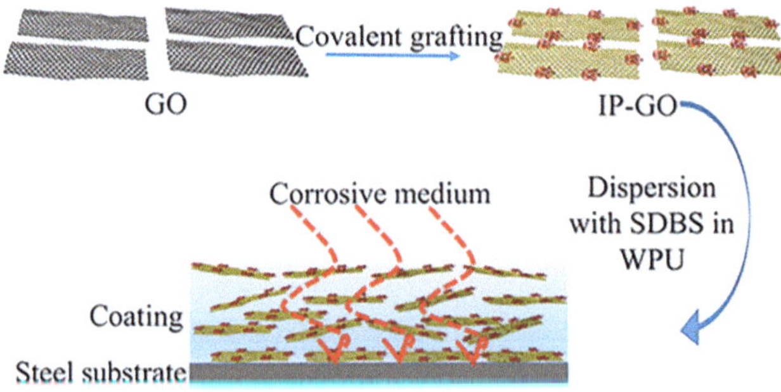

Fig. 5.3 Demonstrating the PU coating functionalized with graphene oxide (GO) causes hindrance between automotive substrate and the corrosive medium [19]

noting how covalent and non-covalent modification of fillers improves dispersion and adhesion. They further discussed active anticorrosion strategies using nanocontainers that release inhibitors into defect zones [18].

5.2.3   Smart Functionality: Self-Healing Coatings

Automotive coatings face constant challenges from mechanical abrasion, environmental exposure, and corrosive agents such as road salt and moisture. Over time, these factors can compromise the integrity of the protective coatings, leading to corrosion, degradation, and aesthetic loss. To address these limitations, recent trend has turned toward smart coating technologies, particularly self-healing polyurethane (PU) coatings. These advanced materials can autonomously repair microdamages and restore their protective functions significantly enhancing durability, reducing maintenance costs, and extending vehicle life cycles. By incorporating self-healing mechanisms, PU coatings evolve from static barriers to dynamic, responsive systems capable of mitigating corrosion even after mechanical damage [20]. This emerging functionality is achieved through strategies like microencapsulation, dynamic covalent bonding, and supramolecular interactions.

Azamian et al. used microcapsules with sodium dodecyl sulfate (SDS) in linseed oil for incorporation in PU coatings. PU coatings containing 4 wt.% of these microcapsules demonstrated over 90% corrosion inhibition after being scratched and exposed to 3.5 wt.% NaCl solution. Another investigation reported 85% healing efficiency and significantly improved corrosion resistance and adhesion when coatings were scratched especially for formulations with 20 wt.% microcapsules, as confirmed via electrochemical impedance spectroscopy and SEM analysis [21]. A composite coating incorporated MXene layers (for barrier effect), CeO_2 corrosion inhibitors, and DA-based self-healing chemistry. It showed efficient sunlight-triggered self-healing with preserved mechanical and anticorrosion properties after repair [22]. Recent development in self-healing PU coatings has explored innovative self-healing formulations tailored for automotive environments, focusing on corrosion resistance, environmental responsiveness, and long-term performance.

5.3   PU Coatings in Aerospace Industries

Polyurethane (PU) coatings have emerged as a critical technology in the aerospace industry, providing advanced surface protection and functional enhancements that are essential for modern aircraft performance and durability. Aerospace environments demand materials that can withstand extreme conditions such as rapid temperature changes, high UV radiation, mechanical abrasion, and exposure to corrosive elements like saltwater and chemicals. PU coatings meet these rigorous requirements through their exceptional mechanical strength, chemical resistance, flexibility, and adaptability to various substrates including metals and composite materials [23].

Introduced initially as protective finishes, PU coatings have evolved significantly with advancements in chemistry and nanotechnology, enabling enhancements in corrosion protection, wear resistance, thermal insulation, and environmental sustainability. They contribute not only to the structural integrity and longevity of aerospace components but also improve weight efficiency and regulatory compliance with reduced hazardous emissions. PU coatings applied on several parts in aeronautical applications. Following are some areas and their function given on which the PU coatings applied.

5.3.1 *Exterior Surface Protection and Aerodynamics*

PU coatings are widely used on aircraft exteriors, including fuselage, wings, landing gear, and engine parts to protect against harsh environmental conditions such as UV radiation, moisture, temperature fluctuations, and abrasion. Modern formulations are engineered to withstand extreme temperatures, weathering, and erosion, which are typical in aerospace operations. These coatings maintain surface integrity and reduce maintenance needs while enhancing the lifespan of aircraft components [24]. PU topcoats are widely used on aircraft exteriors for their exceptional weathering resistance, ability to withstand abrasion, UV exposure, and corrosive environments especially critical at high speeds and in diverse atmospheric conditions. These coatings enhance fuel efficiency by improving aerodynamics and reducing drag [25]. PU coatings also provide excellent barrier protection to aerospace metals, particularly aluminum alloys, safeguarding them against atmospheric corrosion including saltwater exposure. Enhancements are achieved by incorporating nanomaterials such as graphene and halloysite nanotubes, which boost the coatings' barrier properties and durability under mechanical stress and environmental aging. Characterization studies emphasize understanding the microstructure and failure mechanisms to improve the coatings' longevity [26]. Ice formation on aircraft surfaces can seriously degrade performance and safety. Recent research has developed novel PU-based ice phobic coatings that prevent ice adhesion or promote easy ice removal by modifying material parameters and incorporating specialized composites [27]. PU coatings with superhydrophobic and micro-textured surfaces combine fluorinated chemical additives, and particulate fillers reduce surface contamination and insect residue accumulation on aircraft, improving aerodynamics, operational efficiency and delivering durability and low maintenance [28]. PU coatings exhibit excellent adhesion and flexibility on composite aerospace materials, enhancing structural integrity while maintaining light weight. Bio-based and non-isocyanate PU formulations are also being explored to enhance sustainability and tailor thermal and mechanical properties for specific aerospace applications [29, 30]. Specialized PU coatings with UV absorbers and heat-resistant additives are used to maintain coating integrity and appearance throughout long operational cycles.

5.3.2   Primer Topcoat Systems and Structural Substrates

Polyurethane (PU) coatings in aerospace applications also used as primer topcoat systems and are applied over structural substrates to enhance protection, durability, and performance of aircraft components. The PU primer provides strong adhesion to substrates such as aluminum alloys and steel that are typical in aerospace structural components. It primarily offers corrosion resistance and chemical resistance to fluids like hydraulic oils and fuels encountered during aircraft operation. Primers serve as a base to improve bonding of the subsequent topcoat layer. Sometime the PU topcoat adds an external protective barrier against UV radiation, abrasion, moisture, and weathering while providing aesthetic finishes with color stability and gloss retention. It enhances chemical resistance, especially against fluids like Skydrol (hydraulic fluid) and mechanical durability. Collectively the primer topcoat systems are engineered to work synergistically ensuring optimum protection, coating integrity, and repairability. The targeted dry film thickness and formulation ensure fast drying, maintain weldability for metal components, and support easy maintenance [31]. PU coatings adhere well to aerospace materials like metals and composites, maintaining structural integrity under harsh conditions such as temperature changes, moisture, salt spray, and mechanical stress. They offer strong resistance to impact, abrasion, and chemicals, protecting against wear and reducing maintenance needs. It is confirmed by a specific study on aluminum alloy 7075 demonstrates how a PU topcoat over epoxy primers significantly improves wear resistance, cosmetic finish, and anticorrosion performance, with testing confirmed via SEM analysis [32].

Recently Sherwin-Williams' military-grade PU topcoat 85285E has been qualified for extreme weather resistance and chemical and abrasion resistance and meets stringent MIL-PRF-85285E military specifications. It is used as a topcoat combined with compatible primers (e.g., MIL-PRF-23377K epoxy primers) to form durable primer–topcoat systems on military and commercial aircraft [30]. PU topcoat systems have demonstrated very good adhesion on metal and structural steel surfaces, with formulations providing excellent barrier properties against corrosion, UV radiation, moisture, solvents, alkalis, and mechanical impacts [33]. PU primer coatings often impart thermal and chemical resistance, essential for structural parts exposed to fluctuating temperatures and fluids.

5.3.3   Interior Cabin Coatings

Polyurethane (PU) coatings play a crucial role in interior cabin coatings for aerospace applications, providing protective, aesthetic, functional, and safety benefits tailored to the challenging environment inside aircraft cabins. These coatings are engineered to meet stringent aerospace standards, focusing on durability, flame retardancy, and low emissions. These coating plays multiple role in the interior aerospace applications, some of them are given the following:

Protective and aesthetic finishes: PU coatings are widely used on interior plastic and metal components, including overhead bins, seat structures, tray tables, wall panels, lavatories, ceilings, and cockpit parts. They provide a smooth, durable surface finish that resists abrasion, scuffing, and wear from frequent passenger use while maintaining color retention and gloss levels suitable for high-end cabin environments. For instance, two-component PU coating designed for aircraft cabin interiors and cockpits, offering excellent flexibility, hardness, and abrasion resistance with fast drying times, tailored to Boeing material specifications BMS 10-83N Type II and III [34].

Flame retardancy and safety compliance: Interior cabin coatings must meet stringent fire safety regulations. PU coatings are often formulated with flame-retardant additives to achieve compliance with FAA and EASA regulations. These coatings help reduce ignition risks and slow flame spread, protecting passengers and crew in the event of fire [35]. Moreover, aircraft interiors are exposed to frequent cleaning with chemical agents, impacts, and abrasion. PU coatings provide excellent resistance against chemicals, stains, and physical wear, maintaining cabin aesthetics and hygiene over prolonged operational periods.

5.4   PU Coatings in Construction Sector

Synthesis from a chemical tango between polyols and isocyanates, this material doesn't just sit pretty it sticks, flexes, resists, and protects. In the world of construction, where weather, wear, and time are relentless opponents, PU coatings quietly serve as the invisible armor that keeps structures standing strong, looking sharp, and staying safe. PU foams integrated with phase-change materials highlighting PU's versatility and its evolving role in sustainable and high-performance building solutions. This coating use in construction sector for multiple reasons, following are some application areas of the PU coatings in the construction sectors.

5.4.1   Concrete Protection and Waterproofing

PU coatings are widely used to protect and finish concrete surfaces in industrial, commercial, and residential buildings. They serve as sealants and overlays providing moisture resistance, abrasion resistance, and chemical protection on floors and walls subjected to heavy foot and machinery traffic [36]. PU coatings also protect reinforced concrete against chloride ingress, moisture, and chemical attack. Studies show that hydrophobic PU reduces chloride permeability by eight times compared to uncoated specimens, delays crack development, and reduces corrosion rates [37]. These coatings also provide excellent resistance to water penetration, weathering, and mechanical wear, due to their elastic and durable nature. Recently development on waterborne PU coatings with improved biodegradability and environment improved the performance of coating in architectural applications. Recent development occurs in the field of PU coatings, and the properties of PU coatings enhance with magnesium–aluminum layered double hydroxide (PU-LDH) nanofillers (up to 1.5 wt.%). These nanocomposites exhibited improved adhesion strength, impact resistance, abrasion resistance, modulus, and flexibility, alongside enhanced chemical and fire resistance thanks to the thermally stable LDH filler [38].

5.4.2 Architectural and Decorative Finishes

PU coatings are widely used on metal components such as roofing sheets, beams, bridges, storage tanks, piping, and high-exposure architectural surfaces. The coatings enhance resistance against UV radiation, weathering durability, moisture, weathering durability, and pollutants extending the life of metal structures [39, 40]. PU coatings also applied to wood surfaces in construction offer water repellency, scratch resistance, and aesthetic improvements by maintaining gloss and color durability. These coatings also provide protection against microbial attack and weather effects, which are important for both interior and exterior wooden elements [41]. Modern PU decorative finishes aim to combine appealing optics (gloss, transparency, pearlescent effects, and color stability) and functional performance. There develop the UV-curable and waterborne PU formulations tailored for transparent wood with improved flame retardancy and abrasion resistance [41]. High-hardness waterborne polyurethane (WPU) dispersions and two-component (2K) UV-curable WPU systems are designed to produce factory-level gloss with rapid curing and reduced volatile organic compound (VOC) emissions. These formulations are engineered to balance surface hardness and flexibility, preventing cracking while maintaining a high-quality, premium appearance [42].

5.4.3 Flooring and Exterior Wall Coatings

Polyurethane coatings are employed on building facades for their weatherproofing, anti-pollution, and anti-fungal properties. They ensure the structural substrate is protected from rain, UV exposure, and pollutants, reducing maintenance and improving building lifespan. Heavy-duty PU flooring coatings are also used in factories, warehouses, and cleanrooms for abrasion resistance and chemical protection. These systems minimize maintenance and enhance safety. A solid polyurethane coating applied on hardwood floors that provide durability over 8–12 years under normal use with proper maintenance. Real-life use of water-based PU is linked with distinct bead test results to gauge film integrity. They adhere to concrete, wood, masonry, and metal, enhancing surface durability while retaining flexibility to accommodate structural movements. Incorporating nanocellulose into waterborne polyurethane (WPU) coatings significantly enhances their mechanical properties

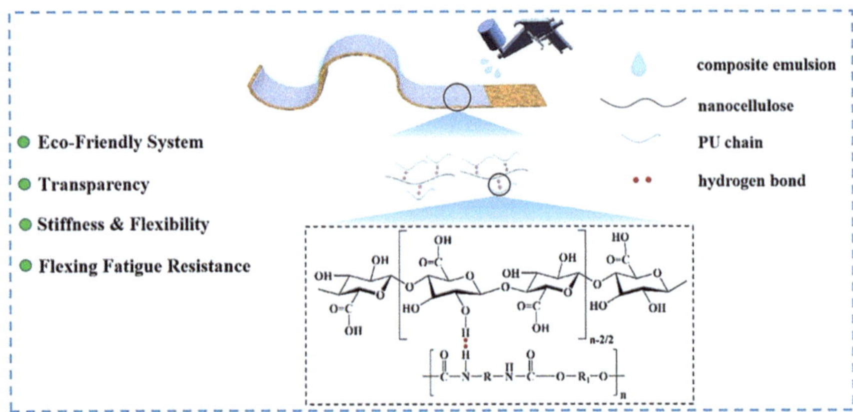

Fig. 5.4 Representing how nanofillers enhance the mechanical properties of PU coating in flooring applications [43]

for the application in cork flooring shown in Fig. 5.4. When poplar wood nanocellulose (WCNF) was added to the WPU emulsion at 0.5 wt.%, the tensile strength increased by 48.58% and the elastic modulus by 118.88%, compared to the pure WPU coating. Similarly, adding bacterial nanocellulose (BCNF) at the same concentration produced even greater improvements, with tensile strength rising by 54.50% and elastic modulus by 120.48% [43].

Bio-based polyurethane (PU) systems incorporating bio-oils, natural fillers, and porous structures can maintain mechanical performance while significantly enhancing thermal and acoustic insulation. The inherent porosity of these materials increases thermal resistance (R-values) and improves sound absorption, contributing to better indoor comfort. Beyond performance, these bio-based formulations reduce environmental impact by lowering volatile organic compound (VOC) emissions and formaldehyde release compared to conventional petroleum-based foams. When further enhanced with phase-change material (PCM) capsules, PU foam composites demonstrate marked improvements in indoor thermal stability, reducing temperature fluctuations and lowering building energy demand. Multiscale modeling confirms their potential to serve as passive solutions for boosting occupant comfort and overall energy efficiency in building envelopes [44].

5.5 PU Coatings in Textile Sectors

Polyurethane (PU) coatings especially waterborne polyurethane dispersions (PUDs) and UV-curable PU hybrids have become a dominant finishing technology for technical and functional textiles because they combine excellent mechanical resilience, chemical and weather resistance, and high formulation flexibility (bio-based polyols, crosslinkers, nanofillers, and additives) while enabling low-VOC processing routes [45]. A bio-based waterborne polyurethane dispersion (PUD) coating, developed for textile applications, was knife-coated onto polyester fabric and characterized by FT-IR and TGA, showing thermal stability above 200 °C and a glass transition temperature of 53.7 °C. The coating achieved elongation above 300%, excellent hydrostatic pressure performance, resistance to QUV aging, hydrolysis, and washing at 40 °C, as well as moderate abrasion resistance. Unlike conventional synthetic polyester PUs, it passed the ISO 15025 fire test without flame-retardant additives and showed reduced bacterial growth on the fabric, though significant antibacterial activity still requires dedicated agents.

In another study, multi-layer flexible devices were developed using graphene as the conductive component and WPU as the protective layer, with knitted fabric substrates produced from chemically recycled waste PET. A wearable textile sensor and Joule heater, fabricated via scratch coating, was optimized at a graphene WPU ratio designated PGW-5. This configuration achieved high conductivity (592 S/m), excellent mechanical performance, and sensitivity while maintaining durability over 2000 stretching–releasing cycles. PGW-5 successfully monitored both large human motions (e.g., elbow flexing) and subtle physiological signals. These properties highlight its potential for health monitoring, wearable electronics, and on-demand thermal therapy applications [45]. The PU coatings are used in various areas and for various purposes in textile sector, some of them given below.

5.5.1   Waterproof Breathable Membranes and Rainwear

Polyurethane (PU) coatings have established themselves as an indispensable finishing technology in the textile sector, owing to their unique balance of mechanical flexibility, chemical resistance, and tunable permeability. One of their most prominent applications lies in the development of waterproof breathable fabrics, where microporous or dense PU films are applied to substrates such as polyester, nylon, or cotton to provide a liquid barrier while permitting moisture vapor transmission [46]. A breathable waterproof layer of PU coating is shown in Fig. 5.5. This dual functionality is critical for high-performance outdoor garments, tents, and protective rainwear, with film microstructure controlled through phase inversion, stretching, or formulation chemistry to achieve the desired hydrostatic head and moisture vapor transmission rate (MVTR) [47]. PU coatings also play a central role in protective clothing and personal protective equipment (PPE), including chemical splash suits and cleanroom garments, where dense, highly crosslinked PU layers enhanced with nanofillers like graphene oxide or nano clay impart solvent resistance, thermal stability, and improved mechanical robustness.

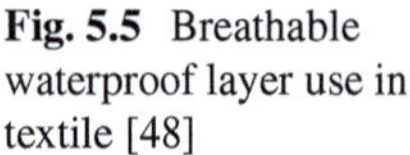

Fig. 5.5 Breathable
waterproof layer use in
textile [48]

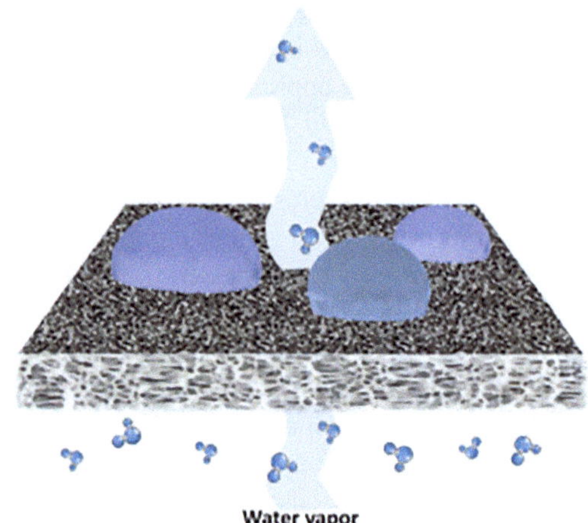

5.5.2   Flame-Retardant and Antimicrobial Textile Finishes

Another major area is flame-retardant textiles, where PU serves as a binder for
halogen-free flame-retardant additives or is chemically modified with phosphorus
and nitrogen containing moieties to promote char formation and slow combustion.
These systems are widely used in upholstery, automotive and aircraft interiors, and
stage drapery, where transparent or low add on formulations are preferred to maintain aesthetics [49]. In the healthcare and hygiene sectors, PU coatings provide
durable antimicrobial properties as shown in Fig. 5.6 through the incorporation or
covalent immobilization of agents such as silver nanoparticles, quaternary ammonium compounds, or organo-selenium molecules, the latter offering particularly
strong wash durability and resistance to microbial colonization [50].

5.5.3   Smart, Conductive, and Sensing Textiles

The emergence of wearable electronics has opened a new field for PU-coated textiles.
By embedding conductive fillers such as graphene, carbon nanotubes, or conductive
polymers within an elastic PU matrix, fabrics can be rendered electrically conductive
while maintaining flexibility, stretch ability, and breathability. Such coatings enable
strain-sensing garments, heated apparel, and integrated wiring for smart textiles, with
waterborne PU systems providing a low-VOC route compatible with delicate fabrics.
PU coatings are equally important in enhancing the mechanical performance of industrial and upholstery fabrics, where they encapsulate fibers, redistribute mechanical
stresses, and improve abrasion and tear resistance, sometimes with reinforcement
from nano-additives like cellulose nanocrystals [50]. Figure 5.7 shows how PU coating is used to create a flexible sensor by dip-coating a sponge with conductive materials, enabling detection of finger bending for wearable sensing applications.

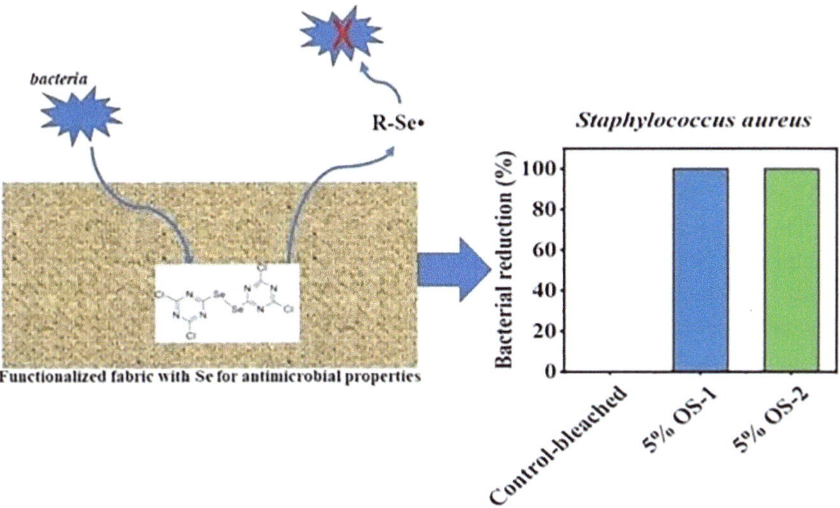

Fig. 5.6 The image illustrates the use of functionalized fabric with selenium (Se) for antimicrobial properties. The fabric interacts with bacteria, leading to bacterial reduction [49]

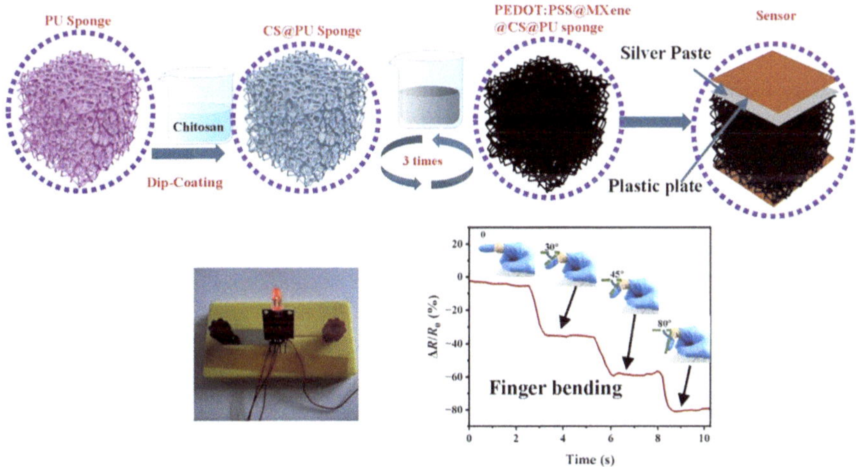

Fig. 5.7 PU coating-based flexible sensor [51]

5.5.4 Aesthetic Finishes, Printing, and Structural Color

In decorative and fashion textiles, PU-based topcoats either waterborne dispersions or UV-curable PU acrylates provide a wide range of finishes, from high gloss to soft matt, as well as excellent adhesion for pigments and printed designs. Finally, PU coatings are applied to filtration and specialty technical textiles, including geotextiles and battery separators, where controlled pore size, surface energy, and chemical resistance are crucial for performance [52]. Across these diverse applications, the ability of PU chemistry to be customized whether for barrier properties, functional responsiveness, or mechanical performance explains its enduring role in textile engineering, with current research focused strongly on bio-based polyols, fluorine-free water repellents, and multifunctional low-VOC systems that meet both performance and sustainability targets.

The integration of polyurethane coatings into textile manufacturing has evolved far beyond simple protective finishes, becoming a multifunctional platform for enhancing durability, comfort, aesthetics, and advanced performance. Recent advancements in bio-based feedstocks, waterborne systems, and nanocomposite reinforcement further align PU-coated textiles with sustainability imperatives while maintaining or even enhancing functional capabilities. This continuous innovation ensures that polyurethane coatings will remain a cornerstone technology in the textile sector, bridging the gap between traditional material needs and emerging demands for intelligent, eco-conscious, and high-performance fabrics.

5.6 Protective Coatings for Metals, Concrete, and Other Surfaces

Polyurethane (PU) coatings are extensively employed as protective layers across a variety of substrates, notably metals, concrete, and diverse engineered surfaces, due to their outstanding mechanical properties, chemical resistance, and versatility. The demand for high-performance protective coatings has accelerated innovations in PU formulations, optimizing their barrier function against corrosion, weathering, mechanical damage, and chemical exposures. Following is the explanation on protective PU coatings tailored for metals, concrete, and other surfaces.

5.6.1 Protection Mechanisms of PU Coatings

PU coatings employ several mechanisms for substrate protection, including physical barriers, chemical passivation, and inhibitor release, and they also provide physical toughness, crack resistance, and self-healing properties. A dense, defect-free PU film creates a physical barrier to oxygen, water, and ions. Incorporating 2D fillers or plate-like inorganic particles increases diffusion path tortuosity, markedly reducing permeation rates. For example, GO-modified WPU shows reduced solvent uptake and improved anticorrosion behavior [53]. Some PU formulations include corrosion inhibitors (e.g., encapsulated inhibitors, ion-exchange molecules) that release at defects to passivate metal surfaces. Recently study demonstrates integration of conductive/active additives (e.g., polythiophene nanoparticles) that increase impedance and lower corrosion current density [54]. Tough, flexible PU networks absorb mechanical stress and resist crack propagation; crosslink density and segment composition (hard vs soft segments) are tuned to balance hardness and toughness. Polyurea/PUs with controlled hard segment content show improved dynamic impact and flexural performance for concrete overlays and metallic substrates [55]. Dynamic covalent chemistries (disulfide exchange, Diels Alder, Schiff base, reversible bonds) and microencapsulated healing agents allow PU coatings to autonomously restore barrier integrity after scratches. Recent primary studies demonstrate thermally or photothermally activated self-healing PUs with restored electrochemical protection and recovered mechanical properties [56].

5.6.2   Protective PU Metals

Metals, including steel, stainless steel, aluminum, and alloys, are prone to corrosion and mechanical wear in harsh environments. PU coatings provide an effective protective barrier, combining adhesion, flexibility, and chemical inertness to extend metal service life.

Anti-corrosive performance of PU coatings achieves through incorporation of nanomaterials and functional additives. As a study exhibits that waterborne PU coatings integrated with Schiff base-functionalized MXene nanosheets demonstrated superior anticorrosion properties due to the synergistic effects of physical barrier and chemical inhibition [57]. PU-polypyrrole composites applied on passivated 316 L stainless steel specimens exhibited improved surface protection against saline corrosion, combining conductive polymer effects with PU's protective. The development of a titanate polyurethane chitosan ternary nanocomposite coating revealed significant improvements in adhesion strength, impact resistance, scratch hardness, and corrosion resistance versus pure PU coatings shown in Fig. 5.8. Titanate nanotubes created an effective lamellar barrier to corrosive species, and chitosan enhanced polymer chain interactions and mechanical properties, making

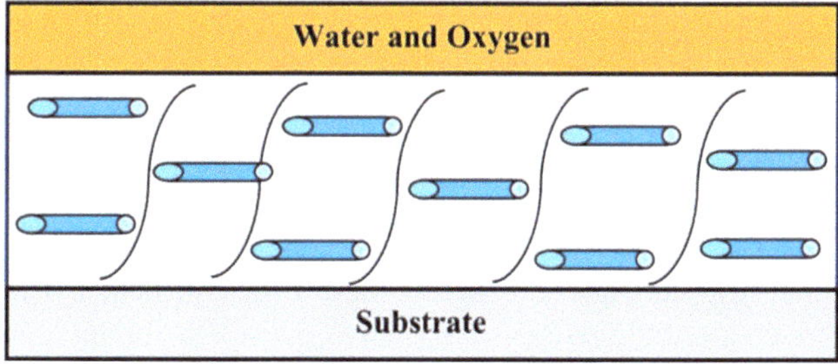

Fig. 5.8 Titanate nano tubes barrier between the oxygen and metal substrate, to slow down the corrosion process [58]

the formulation ideal for metal substrate protection at elevated temperatures and corrosive conditions [58]. Hydrophobic PU coatings are engineered to repel moisture, thereby reducing water ingress and corrosion initiation on metal surfaces. Recent reviews detailed strategies to impart stable hydrophobicity within PU matrices, crucial for applications exposed to humid environments. Multi-layer systems consisting of a primer (epoxy/sol–gel), a midcoat (PU with fillers), and a topcoat (hydrophobic PU or polyurea) are used to obtain long service life. Sol-gel with WPU combinations demonstrated superior salt spray resistance on Al-Zn steels [59]. Figure 5.8 shows how titanate nano tubes become barrier between the oxygen and metal substrate that slow down the corrosion process.

5.7   Protective PU Coatings for Concrete

Concrete infrastructure suffers from environmental degradation due to chemical attacks (chlorides, sulfates), freeze-thaw cycles, and mechanical abrasion. Polymer-modified coatings incorporating PU have emerged as effective protectors by improving adhesion, flexibility, and barrier properties. The integration of PU with silica fume and quartz sand within cement-based coatings markedly improved mechanical properties, durability, freeze-thaw resistance, and sulfate erosion resistance of concrete. Coatings with 6% PU and 12% silica fume demonstrated up to 56% reduction in quality and strength loss after freeze-thaw cycles compared to uncoated concrete or ordinary cementitious [60]. PU-polymers also enhance interfacial bonding performance with concrete substrates, creating flexible protective films that accommodate concrete's microstructural changes without cracking or delamination. A related polyurea coating study emphasized the importance of assessing key determinants for effective concrete protection, highlighting environmental, functional, and executional factors influencing performance and durability in construction projects [55]. Hydrophobic PU reduces water/chloride ingress and therefore delays rebar corrosion; field-oriented studies and modeling indicate significant life extension when used on marine structures [61].

5.8   Protective PU Coatings for Other Surfaces

Beyond metals and concrete, PU coatings are embraced across various substrates such as composites, plastics, wood, and specialized industrial surfaces. PU coatings are adapted with functional fillers and surface treatments to enhance adhesion, mechanical integrity, and environmental resistance on composites and thermoplastics, with research highlighting methods such as plasma treatment to improve coating compatibility. Waterborne PU coatings penetrate and bind well to wood and porous surfaces, combining flexibility and chemical resistance. Modifications include UV stabilizers to prevent discoloration and degradation [55]. PU coatings embedded with natural antifouling agents or hybrid nanocomposites exhibit increased resistance to marine biofouling, protein adhesion, and environmental wear, expanding PU's utility in maritime and biomedical fields. Current trends emphasize self-healing, antimicrobial, and environmentally friendly PU formulations tailored to protect sensitive and variable substrates under complex conditions.

Protective PU coatings have demonstrated immense potential and versatility in safeguarding metals, concrete, and other industrially relevant surfaces. The integration of nanomaterials, polymers blends, hydrophobic modifications, and smart additives has markedly enhanced their performance against corrosion, mechanical damage, environmental aging, and chemical exposures. Concrete-specific polymer-cementitious PU composites show promising improvements in structural durability and resistance to aggressive environmental factors, while metal coatings leverage hybrid nanocomposites and hydrophobic strategies for extended service life. Across other substrates, tailored surface treatments and functional additives sustain PU's long-standing reputation as a premier protective coating material.

5.9   Decorative and Aesthetic Applications

Polyurethane coatings have, over the last decades, evolved from purely protective layers to an advanced design space for decorative and aesthetic finishes across virtually every industrial sector. Their film-forming ability, optical clarity, tunable refractive index, adjustable surface energy, and mechanical behavior, together with multiple curing routes (thermal, UV, dual-cure, waterborne dispersion) make PU systems uniquely suited to realize high-value aesthetic effects while meeting durability requirements. In automotive exteriors, PU clearcoats remain the benchmark for high gloss, depth of color, and long-term gloss retention; modern two-component PU clearcoats are engineered with silane or nanoparticle modifications and controlled crosslink density to resist mar, abrasion, and weathering while preserving optical clarity, and recent work has demonstrated self-healing or disulfide/disulfide exchange chemistries that repair fine scratches and restore appearance after environmental exposure [62, 63]. The chemistry of PU allows for precise control over soft segment selection and molecular weight, which tunes surface modulus. Additionally, controlling hard segment content and crosslink density affects gloss and scratch resistance. Reactive acylation or UV-curable groups enable rapid, high-definition pattern setting for printing applications while minimising thermal exposure to substrates. [64].

In the textile and apparel domain, decorative PU systems extend from pigment and dye printing media to advanced functional aesthetics. UV-curable polyurethane acrylates are the preferred chemistry for high-definition digital and roller printing shown in Fig. 5.9 because they cure rapidly, give excellent color fidelity and gloss control, and minimize thermal damage to substrates [65]. Waterborne PU dispersions produce soft-touch, velvet-like hand and can host colloidal assemblies that create structural colors with reduced environmental burdens versus traditional

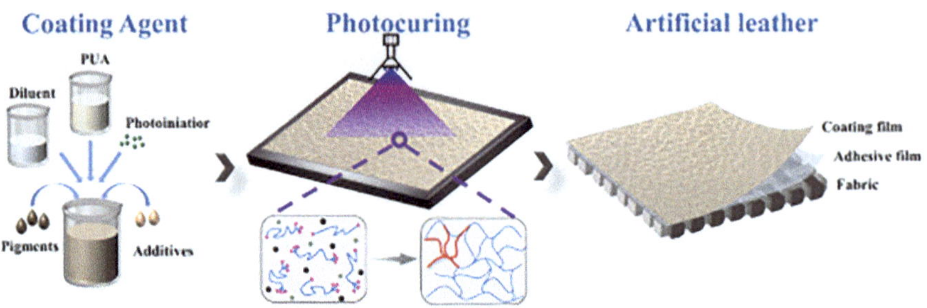

Fig. 5.9 PU coating on fabric to form artificial leather [68]

pigment printing. Thermochromic and photochromic pigments microencapsulated within PU matrices allow garments and soft furnishings to display dynamic color changes while the PU binder protects pigments from laundering and abrasion [11]. Leather, footwear, and synthetic-leather finishes use PU coatings to deliver consistent aesthetic appearance, controlled-gloss, soft-touch, embossed grain replication, and long-term wear. PU lacquers provide a balance of flexibility and film strength that prevents cracking on bending parts (e.g., shoe uppers) while achieving surface effects ranging from full mirror gloss to ultra-matte [66]. PU coatings and photonic/ colloidal combine to generate decorative structural colors on leather substrates while preserving the substrate's hand and flex fatigue behavior [67]. Figure 5.9 shows the formation of artificial leather through PU coating.

Architectural and interior decorative coatings rely on PU technology to realize design intents that combine aesthetics and performance. Self-matting waterborne PU systems give architects and interior designers the soft, contemporary matt finishes increasingly in demand of conversely, high-gloss PU coatings are used for accent panels and high-impact surfaces. Beyond gloss, PU chemistry enables the incorporation of metallic and pearlescent pigments, interference flakes, and even photonic crystal layers to create depth, metallicity, and angle-dependent shimmer with good lightfastness and cleanability [69]. Consumer electronics and appliance exteriors use decorative PU coatings to achieve soft-touch bezels, rubber-like grips, high-gloss control panels, and scratch-resistant glossy finishes. Soft-feel PU create premium haptics while surface-modified nanoparticles or crosslinkers ensure abrasion and solvent resistance. Such coatings can be engineered to suppress fingerprinting or to create tactile contrast between control zones and display areas. Art conservation, restoration, and specialty decorative arts employ PU coatings when clarity, reversibility, and long-term stability are required. PU varnishes have been optimized to provide non-yellowing transparent films for wood and painted artworks; research into photochromic and controlled-gloss PU varnishes supports conservation use where both appearance and aging behavior are critical. Additionally, photonic and structural color approaches consolidated within a PU matrix are being explored for contemporary artists seeking angle-dependent color effects that are robust to handling and exhibition lighting [70].

Finally, the frontier of decorative PU coatings is increasingly multifunctional: decorative layers that also sense, heat, or change color in response to the environment, self-healing clearcoats that restore appearance after scratching, and structural colors preserved under flexible topcoats all point to a future where aesthetics and function are fully integrated. PU coatings are a cornerstone in decorative and aesthetic surface treatments, offering unparalleled flexibility, durability, and visual enhancement combined with protective functions. Recent research indicates continual progress in developing sustainable, multifunctional, and innovative PU coatings that meet the evolving demands of decoration and aesthetics across industries. Their adaptability in finishes, ranging from high gloss to matte, and customizable color range, alongside emerging features such as self-healing and antibacterial effects, ensure PU coatings remain at the forefront of decorative materials technology.

5.10 Emerging Trends and Innovative Uses of Polyurethane Coatings

Polyurethane coatings have long enjoyed a leading role in protective and decorative finishing because of their adaptable chemistry, excellent film properties, and broad processing window. Today, driven by environmental regulation, market demand for multifunctionality, and advances in polymer chemistry and nanotechnology, PU coatings are undergoing a significant renaissance. New generations of waterborne and high solids systems, bio-based feedstocks, non-isocyanate chemistries, UV-curable hybrids, nanocomposite formulations, dynamic/self-healing networks, and multifunctional smart coatings are expanding what PU coatings can accomplish, not just as protective films but as active, sustainable, and design-enabling layers across automotive, aerospace, marine, furniture, textiles, electronics, and specialty art/conservation sectors [71, 72]. The evolution of PU coatings is driven by market demand for high-performance materials that offer durability, aesthetic versatility, and environmental responsibility. The global polyurethane coatings market is rapidly expanding, with projections estimating growth from USD 20.9 billion in 2024 to USD 36 billion by 2034, at a CAGR of approximately 5.6–7% depending on segment and region [73]. The following detailed note synthesizes the current landscape and future directions of PU coatings, suitable for an extended multi-page discussion.

References

1. Srivastava A, Maity S, Das BR (2025) A review on multi-functional polyurethane (PU) coatings for fabric applications: materials, processes and recent developments. Prog Org Coat 207:109377. https://doi.org/10.1016/j.porgcoat.2025.109377

2. Das A, Mahanwar P (2020) A brief discussion on advances in polyurethane applications. Adv Ind Eng Polym Res 3(3):93–101. https://doi.org/10.1016/j.aiepr.2020.07.002
3. Silva TAR, Marques AC, Galhano dos Santos R, Taryba M, Montemor MF (2022) Biobased polyurethane coatings for corrosion protection of carbon steel. Mater Proc 8(1):46. https://doi.org/10.3390/materproc2022008046
4. Polyurethane coatings market size, share & industry analysis, by product type (solvent-borne coatings, waterborne coatings, powder coatings, and radiation-based coatings), by application (automotive & transportation, construction, wood & furniture, aerospace, and others), and regional forecast, 2025–2032. https://www.fortunebusinessinsights.com/polyurethane-coating-market-111428. Accessed 18 Aug 2025
5. A. M. Research Cognitive Market. Polyurethane coating market report 2025. Market size, share, growth, CAGR, forecast, revenue. Cognitive Market Research. https://www.cognitive-marketresearch.com/polyurethane-coating-market-report. Accessed 18 Aug 2025
6. Verma G, Kaushik A, Ghosh AK (2016) Nano-interfaces between clay platelets and polyurethane hard segments in spray coated automotive nanocomposites. Prog Org Coat 99:282–294. https://doi.org/10.1016/j.porgcoat.2016.06.001
7. Polyurethane coatings: a perfect product class for the design of modern automotive clearcoats. ResearchGate. https://www.researchgate.net/publication/329868246_Polyurethane_coatings_a_perfect_product_class_for_the_design_of_modern_automotive_clearcoats. Accessed 18 Aug 2025
8. Delebecq E, Pascault J-P, Boutevin B, Ganachaud F (2013) On the versatility of urethane/urea bonds: reversibility, blocked isocyanate, and non-isocyanate polyurethane. Chem Rev 113(1):80–118. https://doi.org/10.1021/cr300195n
9. Hecking A, Eggert C, Werner U, Yuva N, Behnken G, Reyer R (2018) Polyisocyanate composition based on 1,5-pentamethylene diisocyanate. US20180079852A1. https://patents.google.com/patent/US20180079852A1/en. Accessed 18 Aug 2025
10. Seubert C, Nichols M, Henderson K, Mechtel M, Klimmasch T, Pohl T (2010) The effect of weathering and thermal treatment on the scratch recovery characteristics of clearcoats. J Coat Technol Res 7(2):159–166. https://doi.org/10.1007/s11998-009-9190-4
11. Moon SW, Seo J, Seo J-H, Choi B-H (2021) Scratch properties of clear coat for automotive coating comprising molecular necklace crosslinkers with silane functional groups for various environmental factors. Polymers 13(22):3933. https://doi.org/10.3390/polym13223933
12. Kutschera M, Sander R, Herrmann P, Weckenmann U, Poppe A (2006) Scratch resistance of automobile clearcoats: chemistry and characterization on the micro-and nanoscale. JCT Res 3(2):91–97. https://doi.org/10.1007/s11998-006-0010-9
13. No fear of scratches (2014) PCI Magazine. https://www.pcimag.com/articles/99447-no-fear-of-scratches. Accessed 18 Aug 2025
14. Study on waterborne UV-curable polyurethane acrylate clearcoat and its performance based on nano ZnO as photoinitiator. https://www.google.com/search?q=Study+on+Waterborne+UV-Curable+Polyurethane+Acrylate+Clearcoat+and+Its+performance+Based+on+Nano+ZnO+as+Photoinitiator&rlz=1C1FHFK_en-GBPK1170PK1170&oq=Study+on+Waterborne+UV-Curable+Polyurethane+Acrylate+Clearcoat+and+Its+performance+Based+on+Nano+ZnO+as+Photoinitiator&gs_lcrp=EgZjaHJvbWUqBggAEEUYOzIGCAAQRRg7MgcIARAhGI8CMgcIAhAhGI8C... &sourceid=chrome&ie=UTF-8. Accessed 18 Aug 2025
15. Azamian I, Allahkaram SR, Rezaee S (2022) Autonomous-healing and smart anti-corrosion mechanism of polyurethane embedded with a novel synthesized microcapsule containing sodium dodecyl sulfate as a corrosion inhibitor. RSC Adv 12(22):14299–14314. https://doi.org/10.1039/D2RA01131J
16. Liu Y, Zhan Y, Tian L, Zhao J, Sun J (2024) Study on the anticorrosion and antifouling performance of magnetically responsive self-healing polyurethane coatings. Prog Org Coat 186:108047. https://doi.org/10.1016/j.porgcoat.2023.108047

17. Wen J-G et al (2019) Improvement of corrosion resistance of waterborne polyurethane coatings by covalent and noncovalent grafted graphene oxide nanosheets. ACS Omega 4(23):20265–20274. https://doi.org/10.1021/acsomega.9b02687
18. Salzano de Luna M (2022) Recent trends in waterborne and bio-based polyurethane coatings for corrosion protection. Adv Mater Interfaces 9(11):2101775. https://doi.org/10.1002/admi.202101775
19. Improvement of corrosion resistance of waterborne polyurethane coatings by covalent and noncovalent grafted graphene oxide nanosheets. ResearchGate. https://www.researchgate.net/publication/337356324_Improvement_of_Corrosion_Resistance_of_Waterborne_Polyurethane_Coatings_by_Covalent_and_Noncovalent_Grafted_Graphene_Oxide_Nanosheets. Accessed 22 Aug 2025
20. Chen K et al (2023) Self-healing polyurethane coatings based on dynamic chemical bond synergy under conditions of photothermal response. Chem Eng J 474:145811. https://doi.org/10.1016/j.cej.2023.145811
21. Fathi Fathabadi H, Javidi M (2021) Self-healing and corrosion performance of polyurethane coating containing polyurethane microcapsules. J Coat Technol Res 18(5):1365–1378. https://doi.org/10.1007/s11998-021-00501-0
22. Wang H, Xu J, Du X, Wang H, Cheng X, Du Z (2022) Stretchable and self-healing polyurethane coating with synergistic anticorrosion effect for the corrosion protection of stainless steels. Prog Org Coat 164:106672. https://doi.org/10.1016/j.porgcoat.2021.106672
23. Liu J et al (2019) Degradation behavior and mechanism of polyurethane coating for aerospace application under atmospheric conditions in South China Sea. Prog Org Coat 136:105310. https://doi.org/10.1016/j.porgcoat.2019.105310
24. Polyurethane coatings: a brief overview. ResearchGate. https://www.researchgate.net/publication/283412417_Polyurethane_coatings_A_brief_overview. Accessed 18 Aug 2025
25. Jeeva N, Thirunavukkarasu K, Xavier JR (2024) Influence of multifunctional graphene oxide and silanized vanadium nitride in polyurethane coatings for the protection of aluminium alloy in aerospace industries. Diam Relat Mater 142:110792. https://doi.org/10.1016/j.diamond.2024.110792
26. Jia X, Song X, Fan J, Sun W, Liu C (2022) Research Progress in characterization methods of anti-corrosion and Wear-resistant polyurethane coatings. MATEC Web Conf 358:01045. https://doi.org/10.1051/matecconf/202235801045
27. Bakhshandeh E, Sobhani S, Jafari R, Momen G (2024) From theory to application: innovative ice-phobic polyurethane coatings by studying material parameters, mechanical insights, and additives. Surf Interfaces 51:104545. https://doi.org/10.1016/j.surfin.2024.104545
28. Synthesis and development of polyurethane coatings containing fluorine groups for adhesive applications. T2 Portal. https://technology.nasa.gov/patent/LAR-TOPS-272. Accessed 18 Aug 2025
29. Sherwin-Williams military aerospace coating. Aviation Pros. https://www.aviationpros.com/aircraft-maintenance-technology/mros-repair-shops/paints/painting-equipment-supplies/product/21240332/sherwin-williams-aerospace-coatings-sherwin-williams-military-aerospace-coating. Accessed 18 Aug 2025
30. A two-pack waterborne polyurethane: topcoat for military aircraft. ResearchGate. https://www.researchgate.net/publication/328555110_A_two-pack_waterborne_polyurethane_Topcoat_for_military_aircraft. Accessed 18 Aug 2025
31. Kannan B et al (2025) Waterborne direct-to-metal coating development for anticorrosion coating application on steel. Social Science Research Network, Rochester, NY. https://doi.org/10.2139/ssrn.5191848
32. Vijayakumar T, Senthilvelan T, Venkatakrishnan R (2015) Wear behaviour of polyurethane coated aerospace aluminium alloy (7075). Appl Mech Mater 813–814:252–256. https://doi.org/10.4028/www.scientific.net/AMM.813-814.252

33. Soon LJ, Abdullah TK, Husin ASM, Zubir SA (2025) Enhanced hydrophobicity of polyurethane coating for steel protection through replication and nanomaterial integration. J Thermoplast Compos Mater 38(6):2120–2137. https://doi.org/10.1177/08927057241292303

34. JetFlex® Aircraft Interior Finish. Sherwin-Williams. https://industrial.sherwin-williams.com:443/na/us/en/aerospace/catalog/product/products-by-industry.11543142/jetflex-aircraft-interior-finish.12322086.html. Accessed 18 Aug 2025

35. Yin X, Pang H, Luo Y, Zhang B (2021) Eco-friendly functional two-component flame-retardant waterborne polyurethane coatings: a review. Polym Chem 12(38):5400–5411. https://doi.org/10.1039/D1PY00920F

36. Zafar S, Kahraman R, Shakoor RA (2024) Recent developments and future prospective of polyurethane coatings for corrosion protection—a focused review. Eur Polym J 220:113421. https://doi.org/10.1016/j.eurpolymj.2024.113421

37. Dacuan CN, Abellana VY, Canseco HAR, Cañete MQ (2021) Hydrophobic polyurethane coating against corrosion of reinforced concrete structures exposed to marine environment. Civ Eng Archit 9(3):721–736. https://doi.org/10.13189/cea.2021.090314

38. Guo S, Zhang C, Peng H, Wang W, Liu T (2011) Structural characterization, thermal and mechanical properties of polyurethane/CoAl layered double hydroxide nanocomposites prepared via in situ polymerization. Compos Sci Technol 71(6):791–796. https://doi.org/10.1016/j.compscitech.2010.12.001

39. Wang H, Xu J, Du X, Du Z, Cheng X, Wang H (2021) A self-healing polyurethane-based composite coating with high strength and anti-corrosion properties for metal protection. Compos Part B Eng 225:109273. https://doi.org/10.1016/j.compositesb.2021.109273

40. Li S, Wang S, Du X, Wang H, Cheng X, Du Z (2022) Waterborne polyurethane coating based on tannic acid functionalized Ce-MMT nanocomposites for the corrosion protection of carbon steel. Prog Org Coat 163:106613. https://doi.org/10.1016/j.porgcoat.2021.106613

41. Jian H, Wang H, Wen M, Shi J, Park H-J (2024) High-performance flame-retardant, waterproof, and abrasion-resistant waterborne polyurethane-based transparent coatings for wood protection. Ind Crop Prod 222:119804. https://doi.org/10.1016/j.indcrop.2024.119804

42. Ling Z, Wang H, Zhou Q (2025) 2K UV- and sunlight-curable waterborne polyurethane coating through thiol-ene click reaction. J Compos Sci 9(5):217. https://doi.org/10.3390/jcs9050217

43. Liang G et al (2023) Nanocellulose-reinforced polyurethane as flexible coating for cork floor. Prog Org Coat 178:107480. https://doi.org/10.1016/j.porgcoat.2023.107480

44. Zarmehr SP, Kazemi M, Madasu NGA, Lamanna AJ, Fini EH (2025) Application of bio-based polyurethanes in construction: a state-of-the-art review. Resour Conserv Recycl 212:107906. https://doi.org/10.1016/j.resconrec.2024.107906

45. De Smet D, Wéry M, Uyttendaele W, Vanneste M (2021) Bio-based waterborne PU for durable textile coatings. Polymers 13(23):4229. https://doi.org/10.3390/polym13234229

46. Waterproof breathable polymeric coatings based on polyurethanes. ResearchGate. https://doi.org/10.1177/1528083704045179

47. Cui M et al (2021) A halogen-free, flame retardant, waterborne polyurethane coating based on the synergistic effect of phosphorus and silicon. Prog Org Coat 158:106359. https://doi.org/10.1016/j.porgcoat.2021.106359

48. Tehrani-Bagha AR (2019) Waterproof breathable layers—a review. Adv Colloid Interf Sci 268:114–135. https://doi.org/10.1016/j.cis.2019.03.006

49. Hoque E et al (2023) Antimicrobial coatings for medical textiles via reactive organo-selenium compounds. Molecules 28(17):6381. https://doi.org/10.3390/molecules28176381

50. Zhang Z, Zhang X, Huang W, Zheng X, Ding B, Wang X (2024) Breathable and wearable graphene/waterborne polyurethane coated regenerated polyethylene terephthalate fabrics for motion sensing and thermal therapy. Discov Nano 19(1):61. https://doi.org/10.1186/s11671-024-04004-w

51. Liu M, Liu X, Yang F (2024) Flexible and washable MXene@PEDOT:PSS@CS@PU pressure sensors. Sens Actuators Phys 378:115810. https://doi.org/10.1016/j.sna.2024.115810

52. Sikdar P, Dip TM, Dhar AK, Bhattacharjee M, Hoque MS, Ali SB (2022) Polyurethane (PU) based multifunctional materials: emerging paradigm for functional textiles, smart, and biomedical applications. J Appl Polym Sci 139(38):e52832. https://doi.org/10.1002/app.52832

53. Aramayo MAF et al (2024) Eco-friendly waterborne polyurethane coating modified with ethylenediamine-functionalized graphene oxide for enhanced anticorrosion performance. Molecules 29(17):4163. https://doi.org/10.3390/molecules29174163

54. Ma L et al (2024) Polyurethane coatings with corrosion inhibition and color-fluorescence damage reporting properties based on APhen-grafted carbon dots. Corros Sci 232:112038. https://doi.org/10.1016/j.corsci.2024.112038

55. Waqar A et al (2024) Modeling of success factors of using PU coats in concrete construction projects. Heliyon 10(7):e28908. https://doi.org/10.1016/j.heliyon.2024.e28908

56. Cai Z, Li C, Zhang D, Li J, Gao L (2023) An anti-corrosion coating with self-healing function of polyurethane modified by lipoic acid. J Coat Technol Res 20(2):713–724. https://doi.org/10.1007/s11998-022-00703-0

57. Li X et al (2023) Bio-inspired self-healing MXene/polyurethane coating with superior active/passive anticorrosion performance for Mg alloy. Chem Eng J 454:140187. https://doi.org/10.1016/j.cej.2022.140187

58. Abdellatif AS, Shahien M, El-Saeed AM, Zaki AH (2024) Titanate–polyurethane–chitosan ternary nanocomposite as an efficient coating for steel against corrosion. Sci Rep 14(1):30562. https://doi.org/10.1038/s41598-024-81104-8

59. Yu S-P, Tai H-J (2021) Sol–gel enhanced polyurethane coating for corrosion protection of 55% Al-Zn alloy-coated steel. J Polym Res 28(5):158. https://doi.org/10.1007/s10965-021-02514-0

60. Study on mechanical properties of concrete using silica fume and quartz sand as replacements. ResearchGate. https://www.researchgate.net/publication/325554660_Study_on_mechanical_properties_of_concrete_using_silica_fume_and_quartz_sand_as_replacements. Accessed 18 Aug 2025

61. Zhao H, Wang Q, Shang R, Li S (2025) Development, challenges, and applications of concrete coating technology: exploring paths to enhance durability and standardization. Coatings 15(4):409. https://doi.org/10.3390/coatings15040409

62. Hong JU, Lee TH, Oh D, Paik H, Noh SM (2021) Scratch-healable automotive clearcoats based on disulfide polyacrylate urethane networks. Prog Org Coat 161:106472. https://doi.org/10.1016/j.porgcoat.2021.106472

63. Yu J et al (2025) Fabrication and applications of textile-based structurally colored materials. EcoMat 17:e70012. https://doi.org/10.1002/eom2.70012

64. Chattopadhyay DK, Raju KVSN (2007) Structural engineering of polyurethane coatings for high performance applications. Prog Polym Sci 32(3):352–418. https://doi.org/10.1016/j.progpolymsci.2006.05.003

65. Wang L et al (2022) Amphiphilic alginate stabilized UV-curable polyurethane acrylate as a surface coating to improve the anti-wrinkle performance of cotton fabrics. Prog Org Coat 162:106595. https://doi.org/10.1016/j.porgcoat.2021.106595

66. Soft feel material coatings on the surface of plastic products and their application prospects in the popular fields: a review. https://www.mdpi.com/2079-6412/14/6/748. Accessed 18 Aug 2025

67. Hu J, Han Y, Sun X, Wu W, Huyan J, Wu S (2025) Large-area crack-free photonic crystals coating on leather with brilliant and robust iridescent structural color. Chem Eng J 515:163917. https://doi.org/10.1016/j.cej.2025.163917

68. Ma J et al (2022) Synthesis and properties of photocurable polyurethane acrylate for textile artificial leather. Prog Org Coat 171:107017. https://doi.org/10.1016/j.porgcoat.2022.107017

69. Jin Q et al (2025) Self-matting waterborne polyurethane coatings with ultra-low gloss and enhanced corrosion resistance via molecular design and ZnO integration. Polym Adv Technol 36:e70268. https://doi.org/10.1002/pat.70268

70. Construction of rainbow-like structural color coatings on wood surfaces based on polystyrene microspheres. https://www.mdpi.com/1999-4907/14/1/76. Accessed 18 Aug 2025

71. Gavande V, Mahalingam S, Kim J, Lee W-K (2024) Optimization of UV-curable polyurethane acrylate coatings with hexagonal boron nitride (hBN) for improved mechanical and adhesive properties. Polymers 16(17):2544. https://doi.org/10.3390/polym16172544
72. Malucelli G, Lorenzetti A (2025) Sustainability in polyurethanes: old hat or new strategy for future developments? Npj Mater Sustain 3(1):20. https://doi.org/10.1038/s44296-025-00064-w
73. Polyurethane coatings market analysis—China, US, Japan, Germany, UK—size and forecast 2024–2028. https://www.technavio.com/report/polyurethane-coatings-market-industry-analysis. Accessed 18 Aug 2025

Chapter 6
Environmental Impact and Sustainability of PU Coatings

6.1 Need for Sustainability

Sustainability has emerged as a central focus in materials science, especially for polymeric coatings like polyurethanes (PU), which play a critical role across diverse sectors, including automotive, aerospace, marine, construction, textiles, wood/furniture, packaging, and electronics. Traditionally prized for their mechanical robustness, chemical resistance, superior film formation, and aesthetic versatility, PU coatings now face growing environmental concerns, regulatory pressures, and consumer demand for eco-friendly solutions. This calls for a paradigm shift toward more sustainable PU coatings that minimize environmental impact throughout their entire lifecycle from raw material sourcing, production, and application to in-service performance and end-of-life management without compromising on quality or durability. Traditional solvent-borne and isocyanate-based systems are under increased scrutiny due to volatile organic compound (VOC) emissions, worker safety risks, and challenges in disposal. In response, recent research highlights a clear transition toward lower-emission, waterborne PU formulations, incorporation of bio-based feedstocks, non-isocyanate synthesis routes, and innovative design-for-recycling concepts such as dynamic covalent networks and vitrimers. Importantly, sustainability in PU coatings is recognized as a multi-criteria optimization problem, requiring a careful balance of performance, scalability, cost-effectiveness, and circularity to meet the evolving demands of industry and society [1, 2].

Z. Zubair et al., *Functional Polyurethane Coatings*, SpringerBriefs in Materials, https://doi.org/10.1007/978-3-032-12730-3_6

6.2 Regulatory and Market Drivers for Sustainable PU Coatings

Globally, environmental regulations such as the European Union's "REACH," US EPA limits on air pollutants, and similar mandates in North America and Asia-Pacific are accelerating the phase-out of high-VOC coatings and toxic chemicals used in PU manufacture. This has spurred investment in waterborne, solvent-free, and bio-based PU technologies. Policy drivers include VOC and hazardous air pollutant limits, circular-economy directives, and corporate GHG commitments. Market forces amplify these pressures, and OEMs and retailers specify low-VOC finishes and require documentation of renewable content and LCA. Recent studies observe rapid growth in commercial waterborne systems with still-uneven parity versus solvent-borne benchmarks, especially in extreme environments. Market awareness and consumer preferences align with regulatory trends, as industries including automotive, construction, and packaging seek coatings that combine sustainability with durability and aesthetic excellence. According to recent market forecasts, the solvent-free PU coatings segment is expected to grow at a CAGR of 7% from 2025 to 2033, propelled by environmental compliance and performance demands [3].

6.3 Bio-based and Renewable Raw Materials in PU Coatings

A major stride in sustainable PU coatings involves substituting fossil-based polyols with bio-based alternatives. Natural oils such as castor, linseed, soybean, and particularly camelina oil have emerged as functional polyol precursors. These bio-sourced materials offer advantages, including renewability, biodegradability, and a lower carbon footprint [4]. Researchers have synthesized polyester amide polyols from camelina oil, demonstrating comparable or superior film properties relative to petrochemical counterparts. Incorporation of lignin, tannins, and succinic acid derivatives further enhances the sustainability profile while imparting unique mechanical and thermal characteristics, as shown in Fig. 6.1. These developments encourage the shift from fossil feedstocks to circular, biologically derived chemistries. Advances include tailored functionality for crosslink density, improved hydrolysis resistance, and compatibility with waterborne systems. Multiple studies report bio-based PUs meeting abrasion and chemical-resistance targets in wood, packaging, and construction coatings [5]. Key challenges remain supply-chain consistency, color, and oxidation stability, but the trajectory is positive and supported by LCAs that quantify benefits under realistic energy scenarios [6].

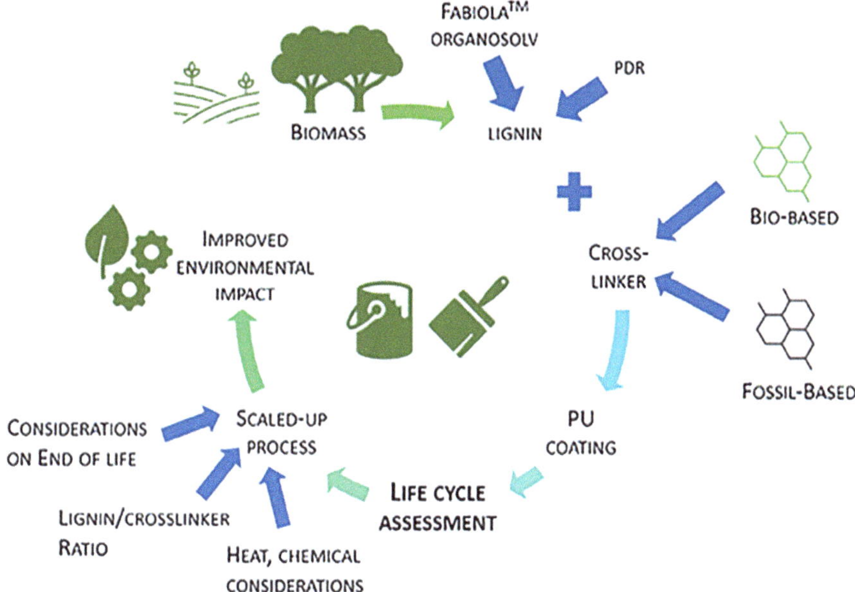

Fig. 6.1 A concise lifecycle view of bio-based PU coatings, biomass feedstocks enable bio cross-linking with lignin, blended with fossil components to form coatings, with life cycle assessment guiding environmental impact, scale-up, and end-of-life consideration [6]

6.4 Innovations in Non-isocyanate Polyurethane (NIPU) Coatings

Innovations in non-isocyanate polyurethane (NIPU) coatings have been making significant strides recently, offering sustainable, safer, and highly functional alternatives to conventional polyurethane coatings that rely on toxic isocyanates for synthesis. NIPU coatings are primarily synthesized through the aminolysis reaction of cyclic carbonates (CCs) with polyfunctional amines, resulting in polyhydroxy urethane (PHU) structures without using harmful isocyanates. This approach reduces toxic exposure risks and environmental impact while maintaining key polyurethane properties [7]. Cyclic carbonates can be sustainably obtained from renewable feedstocks, such as epoxidized vegetable oils (e.g., sunflower, linseed, cardanol), reacted with CO_2, contributing to a circular bio-economy approach to NIPU production. Alternative routes include transurethanization of bis-alkyl carbamates with alcohols, offering solvent-free and catalyst-free options for eco-friendly polymer synthesis [2].

The amino-telechelic NIPU approach allows precise control over polymer soft and hard segment distribution, enabling tuning of flexibility, hardness, thermal stability, and chemical resistance for diverse performance requirements. This fine-tuning is particularly useful for room temperature curing coatings that meet stringent industrial standards, such as aviation coatings with excellent low-temperature flexibility down to −54 °C [8].

Two-component (2K) and one-component (1K) waterborne NIPU coating systems have been developed, allowing ambient curing and superior environmental resistance, broadening application scopes to industrial coatings. NIPU foams produced by combining cyclic carbonates and polyamines have been optimized using blowing agents or self-blowing methods, creating lightweight, thermally insulating, and mechanically robust foam products suitable for packaging, insulation, biomedical scaffolds, and lightweight automotive/aerospace materials [9]. These foams demonstrate attractive, eco-friendly characteristics compared to traditional PU foams made from isocyanates. Morphological and mechanical characterization studies inform optimization of foam density, porosity, elasticity, and thermal stability tailored for end performance.

Since no isocyanates are used, NIPU coatings avoid the respiratory toxicity and sensitization risks associated with conventional PU manufacturing and processing, improving workplace safety and regulatory compliance. The absence of volatile organic compounds (VOCs) and reduced energy needs for curing due to room temperature polymerization further enhance environmental benefits and reduce operational costs. Innovations in non-isocyanate polyurethane coatings are rapidly advancing through sustainable synthetic routes, bio-based monomers, molecular-level property control, and hybrid chemistry development. These coatings offer safer, environmentally friendly alternatives to traditional PU coatings while matching or exceeding performance criteria for industrial, biomedical, and advanced material applications.

6.5   Solvent-Free, Waterborne, and Low-VOC Polyurethane Coatings

Solvent-free, waterborne, and low-VOC polyurethane (PU) coatings are key eco-friendly advancements in the coatings industry that address environmental and health concerns while maintaining performance.

Solvent-free PU coatings are formulated without organic solvents, eliminating volatile organic compound emissions during application and curing. They usually involve a two-component system (polyol and polyisocyanate) mixed just before use. The lack of solvents leads to high solids content, enabling thicker coats with fewer applications. Solvent-free PU coatings are prepared by mixing carefully proportioned polyols (e.g., polyether or polyester polyols) and polyisocyanate components under controlled conditions [10]. Fillers such as glass flakes, silica powders, and titanium dioxide are often included for mechanical strength and durability. The reaction occurs at room temperature with no need for high temperatures or solvents, improving safety and reducing energy use [11]. These coatings exhibit excellent mechanical properties, including impact resistance, flexibility (anti-cracking under bending), salt spray resistance, and cathodic disbonding resistance, even in harsh environments like offshore structures and pipelines. The balanced formulation ensures good flow and leveling without solvents.

Waterborne PU coatings use water as the primary dispersion medium instead of organic solvents, offering a substantial reduction in VOCs while maintaining coating properties. Waterborne PU dispersions can be made via prepolymer or acetone processes, where the hydrophilic groups are introduced to the polymer chain to stabilize the dispersion in water. Semi- or full waterborne formulations allow UV-curable or ambient-cure coatings with good film formation [12]. They provide good adhesion, flexibility, and chemical resistance with environmental benefits. Challenges include achieving good flow and film homogeneity due to water evaporation and maintaining long-term durability. Recent advances in UV-curable waterborne PU acrylates provide solvent-free, environmentally friendly coatings with excellent hardness and chemical resistance [4]. Low-VOC PU coatings contain significantly reduced amounts of organic solvents compared to traditional solvent-based PU coatings, without eliminating the solvent in many cases. Utilizing solvent-free polyol components and reactive diluents, such coatings reduce VOC content while maintaining ease of application. Some waterborne PU coatings still require small amounts of co-solvent (<10%) to ensure film formation and film integrity. These coatings support regulatory compliance with environmental laws and reduce health hazards from inhalation of solvents while maintaining good mechanical and protective properties [13].

6.6   Sustainable Additive Engineering and Multifunctionality

Additive engineering and multifunctionality play a crucial role in enhancing the sustainability of polyurethane (PU) coatings by enabling the integration of environmentally friendly properties, improved performance, and extended durability. Through the strategic incorporation of advanced additives, PU coatings become more sustainable while maintaining or even enhancing their mechanical, thermal, and protective qualities [14].

Incorporating nanomaterials such as titanium dioxide (TiO_2), polyaniline, halloysite nanotubes, carbon nanotubes, and other nanostructures into PU coatings improves corrosion resistance, self-cleaning ability, and environmental protection of coated surfaces. These additives boost the coating's durability and functionality without needing excessive chemical inputs or solvent-based additives [15]. For example, PU coatings combined with a nanocomposite of TiO_2/polyaniline/halloysite/carbon nanotubes have shown significantly higher corrosion resistance and self-cleaning capabilities, which reduces maintenance and replacement frequency, promoting sustainability [16]. Additives derived from renewable biomaterials, particularly lignin and its modified forms, are increasingly used to replace or supplement petroleum-based components in PU coatings. Lignin, a natural polymer abundant in wood, offers hydroxyl groups that can be chemically tailored to enhance crosslinking, mechanical strength, flame retardancy, UV protection, and hydrophobicity in PU coatings [17]. Partial depolymerization and fractionation of lignin enable selective modification of its molecular weight and reactivity, allowing precise tuning of coating properties like flexibility and thermal stability while significantly increasing biomass content (90%) in coatings. This strategy substantially reduces reliance on fossil resources and contributes to circular bio-economy goals [18].

Additives can impart multiple functionalities to PU coatings beyond traditional protection, such as UV blocking, antimicrobial activity, flame retardancy, and photocatalytic self-cleaning. For instance, incorporating nanoparticles that absorb UV or generate reactive oxygen species under sunlight helps prevent degradation of the coating and underlying substrate, enhancing longevity and reducing environmental impact. Similarly, additives that confer hydrophobic or oleophobic properties help reduce cleaning chemicals and water use, boosting sustainability. Combining different nanostructured additives in PU matrices can lead to synergistic effects that optimize mechanical toughness and corrosion resistance. This reduces coating failure rates and prolongs service life, key to reducing material waste and resource consumption over time [19]. Additive engineering systematically enhances the sustainability of PU coatings by incorporating advanced nanomaterials and bio-based compounds that improve durability, reduce environmental impact, and introduce multifunctional protective properties. These innovations enable the development of next-generation PU coatings aligned with environmental regulations and sustainability goals in the coatings industry.

6.7 Circularity (Vitrimers, Reprocessing, and Chemical Recycling)

Polyurethane (PU) coatings, widely used due to their durability, chemical resistance, and versatility, face sustainability challenges rooted in their thermoset nature, which traditionally precludes easy recycling or reprocessing. Circularity in PU coatings aims to close the material loop by enabling the reuse, reprocessing, or regeneration of PU materials to reduce waste and environmental impact. This entails developing materials and processes that allow PU coatings to be recycled chemically or physically, providing pathways toward a sustainable lifecycle consistent with circular economy principles. PU coatings are often thermoset polymers with a crosslinked three-dimensional network that imparts desirable mechanical strength and chemical stability but prevents reshaping or conventional recycling via melting. This crosslinked network hinders mechanical recycling, and disposal alternatives like landfilling or incineration pose environmental hazards. Therefore, achieving circularity in PU coatings requires innovative chemistries and recycling technologies that can break or reorganize the polymer network without significant degradation of material properties [20].

Vitrimers are a class of covalent adaptable networks (CANs) with dynamic covalent bonds that can undergo thermally induced exchange reactions while maintaining crosslink density. This dual behavior allows thermoset-like performance at use temperature and flow or reshaping under heat, enabling repair, recycling, or reshaping of PU coatings. Vitrimer bonds resist flow at room temperature but relax under elevated temperatures. In the field of PU coatings, Li et al. first proposed a PU vitrimer with carbamate exchange chemistry enabling strong mechanical properties and re-processability without significant property loss [21]. A notable advancement is the bio-based, catalyst-free PU vitrimer developed as shown in Fig. 6.2, featuring strong tensile strength (76.8 MPa), excellent impact energy absorption, and, importantly, thermal repairability at 160 °C in just 10 min, while fragmented surfaces were fully reformed via hot pressing at 180 °C within 2 h. This work demonstrates lab-scale feasibility for re-smelting and repairing PU coatings without catalysts a major technical breakthrough [22].

These vitrimer PU coatings combine the performance of thermosets with the recyclability of thermoplastics, providing a sustainable circular solution that permits multiple reuse cycles with reduced waste. Beyond coatings, vitrimers have found broader application in flexible electronics, such as recyclable printed circuit boards using transesterification vitrimers. These printed boards can be repaired multiple times and dissolved without destructive solvent use, achieving 98% polymer recovery highlighting the promise of vitrimer technology for durable, circular electronic components [23].

Chemical recycling focuses on depollution and depolymerization of PU materials back into monomers or valuable precursors for producing new PU coatings. This contrasts with mechanical recycling (grinding, reuse) by targeting molecular-level breakdown to enable regeneration of virgin-quality chemicals. Glycolysis, Hydrolysis, aminolysis, acidolysis, and phosphorolysis are some traditional chemical recycling methods that cleave urethane and ester bonds to yield polyols, amines, or acids for reuse. These methods often need harsh conditions like high pressure and temperature [24]. A recent breakthrough uses an organoboron Lewis acid catalyst to

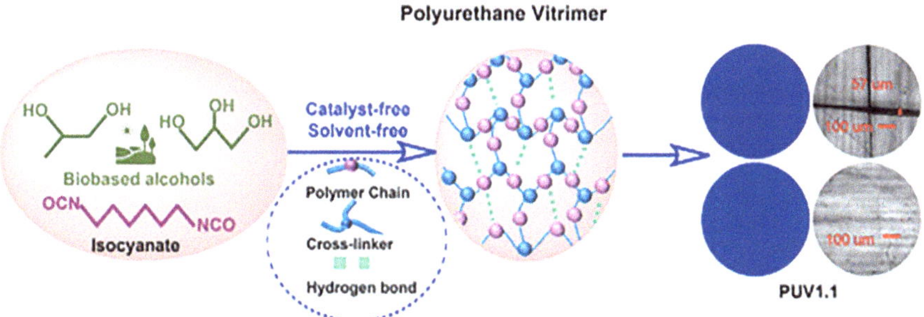

Fig. 6.2 Formation of a polyurethane (PU) vitrimer via condensation of bio-based reactants and catalysis-free exchange to yield a crosslinked, recyclable polymer network. The left panel shows bio-based alcohols and isocyanate reacting to form a polyurethane linkage [22]

depolymerize thermoplastic and thermoset PUs at mild temperatures (80 °C) in toluene, selectively regenerating isocyanates like MDI and HDI without hazardous reagents such as phosgene shown in Fig. 6.3. The recovered isocyanates were successfully re-polymerized with polyols to produce second-generation PUs with properties comparable to virgin materials, representing a scalable chemical recycling path toward PU circularity [25].

For PU coatings, mechanical recycling typically offers limited value; powdered PU serves as a filler rather than a regenerative coating. Still, designs that facilitate chemical re-powdering and reprocessing under vitrimer chemistry or dual-cure systems enable repair or re-layering, making refurbishment viable without discarding the substrate. Covalent adaptable networks (CANs) with disulfide bonds in PU foams exhibit malleability and reprocess ability akin to thermoplastics. Disulfide exchange mechanisms enable multiple reprocessing cycles with minimal loss in performance, particularly in PU foams and coatings, where enhanced repair capacity extends lifetime and supports circular life data.

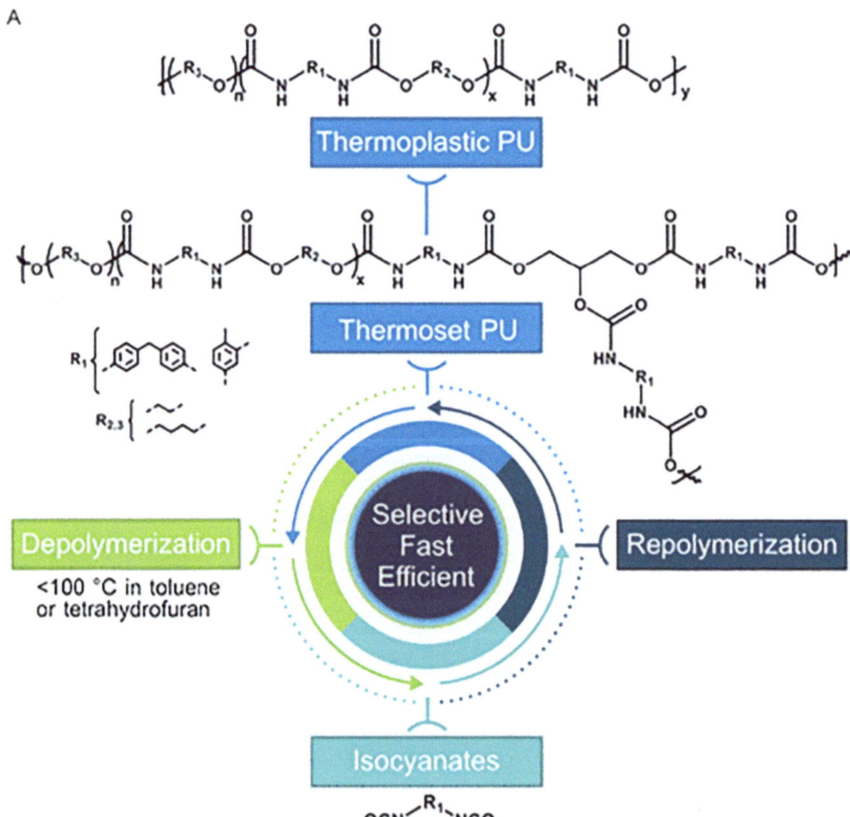

Fig. 6.3 The elaboration of polyurethane depolymerization into isocyanates under mild conditions, followed by their subsequent repolymerization into second-generation materials [25]

6.8   Handling and Storage Guidelines for Polyurethane Coatings

Safeguarding PU coatings through appropriate storage and handling ensures not only occupational safety but also the maintenance of coating properties such as adhesion, color, and durability. This is vital for the end-use industries, including construction, automotive, aerospace, and textiles, where PU coatings contribute to product performance and lifecycle sustainability. PU coatings are widely adopted across industries for their versatility, durability, and resistance to harsh environments. Due to their chemical complexity and broad performance range, safe and effective handling and storage practices are essential for preserving product quality, ensuring personnel safety, and meeting regulatory requirements. Followings are some handling and storage guidelines for PU coatings that must be kept in mind when you deal with the PU coating in different applications.

6.8.1   General Storage Principles and Environment for PU Coatings

Store finished coatings and intermediates in tightly sealed containers under cool, dry, and well-ventilated conditions, away from ignition sources and direct sunlight. Temperature excursions accelerate side reactions and destabilize dispersions via viscosity drift or coagulation. Study on waterborne dispersions emphasizes that stable particle morphology depends on ionic content, counter-ion choice, and pH; storage outside recommended ranges undermines colloidal stability and film formation [26]. Use clean, dry, inert headspaces for moisture-sensitive components and avoid reactive metals for storage hardware. Rotate stock on a first-in-first-out (FIFO) basis and record batch temperatures and visual checks [27].

PU coatings should be stored in areas maintained between 10 and 30 °C and never exposed to freezing or excessive heat. Most manufacturers specify a preferred storage temperature range (less than 25 °C) to ensure long-term stability and avoid premature curing or degradation. Excess humidity induces hydrolysis and may degrade the polymer chains, leading to loss of coating quality and shorter shelf life. A well-ventilated, dry environment minimizes condensation and microbial growth, preserving integrity. Adequate airflow allows the off-gassing of volatile compounds, particularly from waterborne and solvent-based formulations, mitigating the risk of VOC buildup [12].

6.8.2   Safe Containment, Isolation, and Labeling

Safe containment, isolation, and labeling of polyurethane (PU) coatings are critical practices to protect people, the environment, and facilities at all stages from manufacturing and storage to application and disposal. PU coatings must be stored in secure, tightly sealed containers made of compatible materials to prevent leaks, spills, and environmental contamination. Facilities should use appropriate containment such as bunds, spill control kits, and dedicated storage rooms separated from other chemicals. Use containment systems to prevent release into the environment. Storage areas should be equipped with ventilation, fire suppression, and spill management infrastructure especially for isocyanates and polyols, which can present specific hazards [28]. Maintain recommended temperatures to prevent degradation. Pressurized containers should never be tampered with before being properly depressurized. Connections and interchange lines must be clearly labeled and equipped with color-coded markings to prevent mishandling, spills, and cross-contamination in tank farms [29].

When isolating PU coatings (e.g., between manufacturing steps or in storage), ensure physical separation from incompatible substances such as strong acids, bases, or oxidizers. Best practice is to have separate, fire-resistant compartments for isocyanates and polyols, limiting fire spread, cross-contamination, and leakage [30]. Storage facilities should implement fire detection and suppression systems throughout all compartments. Isolate work areas for PU handling and ensure proper ventilation to minimize exposure to vapors. Personnel must use personal protective equipment (PPE) such as chemical-resistant gloves, safety goggles, respirators, and protective clothing, especially during mixing, spraying, and application processes.

6.8.3   Shelf Life and Stability

In typical warehouse conditions below 25 °C and out of sunlight, shelf life ranges between 12 and 24 months depending on the formulation. Periodic testing (viscosity, color, fineness) is recommended post-storage to ensure the coating's stability. Loss of gloss, discoloration, excessive thickening, or crystallization are signs of compromised storage stability and should prompt disposal or remediation [31]. PUA resins and formulated inks must be protected from actinic light (UV and high-energy visible) during storage and handling; even brief exposure can initiate polymerization. Use amber containers, light-blocking films, and low-UV lighting in staging areas [32]. Oxygen inhibition leads to tacky surfaces or under-cure; recent studies explore photo-initiator selection and photoinduced electron transfer (PET) strategies to suppress inhibition and improve through-cure, which is informative for selecting and storing compatible initiator packages [33]. Do not refrigerate below manufacturer limits because certain photoinitiators crystallize and separate. Keep containers tightly closed to prevent oxygen ingress that depletes inhibitors. Pre-mix under low-light, use light-shielded pumps/hoses, and stage filled applicators with opaque covers. Where oxygen ingress is likely (open trays, gravure cells), consider inert gas blankets and fast-cure workflows validated by draw-downs and real-time FTIR.

WPU/PUD storage stability hinges on particle morphology, ionic stabilization, solids content, and neutralization level. Recent experimental work emphasize that optimized hard-segment content, neutralizer type, and surfactant selection improve resistance to sedimentation, viscosity drift, and freeze–thaw cycling. Maintain storage at 5–35 °C; avoid freezing unless explicitly rated "freeze–thaw stable," as ice crystallization can rupture particles and cause irreversible coagulation. For shipments through cold chains, specify freeze–thaw cycles per ASTM D2243 and verify post-cycle particle size and viscosity before release. Prolonged storage can shift pH, alter zeta potential, and trigger flocculation. Recent morphology-control studies demonstrate that counter-ion choice and degree of neutralization affect particle packing and film coalescence, with direct implications for shelf life [34].

6.8.4   Handling Practices

Handling and safety practices for polyurethane (PU) coatings are essential to protect workers, maintain product quality, and minimize environmental and health risks. Always handle PU coatings in open or well-ventilated areas to prevent the accumulation of harmful vapors and reduce inhalation risk. Local exhaust ventilation or fume extraction systems should be used where possible to capture airborne contaminants. Avoid closed or confined spaces unless equipped with proper respiratory protection and ventilation systems. Wear appropriate PPE, including chemical-resistant gloves, safety goggles or face shields, long-sleeve clothing, and respirators when necessary. Use gloves made from materials resistant to PU components to prevent skin contact, as PU can cause skin irritation or sensitization [35]. Respirators should be selected based on the exposure level and specific isocyanates or solvents present. Store PU coatings in tightly sealed, labeled containers in cool, dry, and well-ventilated areas away from heat, sparks, and open flames to prevent fire hazards. Keep away from incompatible chemicals and sources of ignition. Prevent spills and leaks by using secondary containment and routine inspections of storage areas. Do not smoke or use ignition sources in handling areas as PU coatings are flammable.

Follow manufacturers' recommendations for mixing ratios and curing conditions to ensure proper chemical reaction and performance. Avoid direct contact with wet coatings; clean equipment immediately using recommended solvents to minimize contamination and skin exposure. Use application techniques such as spraying, brushing, or rolling carefully to control airborne particles and overspray. Wash hands thoroughly after handling PU coatings and before eating, drinking, or smoking to prevent ingestion or skin contamination. Remove contaminated clothing promptly and avoid prolonged skin contact. Barrier creams may help but should not replace gloves. In case of spills, use appropriate absorbents and dispose of waste in compliance with local regulations. Avoid release to water or soil [36]. Train all personnel in safe handling, emergency response, and correct use of PPE. Maintain accessible Material Safety Data Sheets (MSDS) and emergency contact numbers at workplaces. Have fire extinguishers (foam, CO_2, dry chemical) and first-aid supplies readily available.

6.8.5   *Health and Safety Precautions for Workers*

Polyurethane coatings deliver excellent durability, chemical resistance, and aesthetics, but many popular systems cure via reactions of diisocyanates with polyols. Unreacted NCO groups can be present in mists, vapors, and on surfaces during mixing, spraying, and early curing. These exposures are strongly associated with respiratory and dermal sensitization, occupational asthma, and dermatitis; once sensitized, even trace exposures can trigger severe symptoms. Recent studies highlight the high toxicity and sensitization potential of isocyanates used in coatings and similar applications [37]. In addition to isocyanates, PU coatings may contain solvents, amine catalysts, and additives that present their own hazards (CNS depression, irritation, flammability). Contemporary occupational reviews specific to paint workers still flag this combined hazard mix and the need for layered controls [38]. The spray application is the dominant inhalation route, with short high-intensity peaks of HDI/IPDI iso-cyanurates and monomers in painters' breathing zones. Even in modern shops, short-term exposures during spraying and cleaning/maintenance of spray equipment can be significant [39]. Dermal exposure from an afterthought to a primary concern. Evidence (human and animal) shows skin exposure alone can induce systemic sensitization and subsequently precipitate asthma after later inhalational challenges, meaning perfect respirator use is not enough if skin is unprotected [40].

Workers face additional risks like musculoskeletal injuries from handling large containers and repetitive spray application, Fire and explosion hazards due to the flammable nature of many PU coatings and solvents, and noise exposure from industrial painting processes. Among booth types, downdraft spray booths consistently achieve the lowest overspray and isocyanate concentrations near painters when compared with cross draft or semi-downdraft designs. NIOSH and EPA compilations, as well as collision-repair field studies, show orders-of-magnitude reductions in particulates/overspray and measurable decreases in isocyanate exposure with downdraft airflow and HVLP spray technology [41]. Supplied-air respirators (SAR) are preferred for spray application of isocyanate-containing coatings especially in high-load or small-booth conditions because half-mask APRs may be inadequate during intense overspray peaks, and cartridge warning properties are poor. OSHA/NIOSH documents and field investigations align on this. If using air-purifying respirators, use full-face (or PAPR) with appropriate filters only after exposure assessment verifies adequacy and with a robust change-out schedule; ensure quantitative fit testing and booth downdraft.

For workers handling polyurethane coatings, effective personal protective equipment is essential to minimize exposure to hazardous chemicals. Gloves made of butyl rubber offer the best resistance to HDI and IPDI, while nitrile gloves, often preferred over natural latex, provide practical protection, though breakthrough can still occur, so double-gloving and regular replacement are advisable [42]. Using laminate films like PE/EVAL as inner barrier layers under durable gloves further enhances safety. Disposable coveralls and sleeves made from polypropylene/

polyethylene laminates provide consistent protection against isocyanate permeation, unlike cotton garments, which absorb coatings and are not protective [43]. Eye and face protection should include full-face respirators or tight-fitting goggles with face shields to prevent splashes and aerosol contact with the eyes and skin, ensuring comprehensive defense during PU coating tasks. For PU coatings, real safety comes from substitution, enclosure, ventilation, disciplined PPE, and biomonitoring, not from any one element. The recent innovation on dermal pathways is decisive, booth performance is pivotal, and competency-based training sets the new baseline. Build your program around these pillars and verify it with task-based air sampling and biological monitoring. That's how you keep painters healthy while meeting modern finish-quality demands. The safety of workers in PU coatings depends on a multidisciplinary and proactive approach integrating up-to-date research findings, comprehensive PPE, engineering controls, training, and robust health monitoring. Using safer chemical alternatives and ongoing education will lower risks and protect worker health.

References

1. Pieters K, Mekonnen TH (2024) Progress in waterborne polymer dispersions for coating applications: commercialized systems and new trends. RSC Sustain 2:3704–3729. https://doi.org/10.1039/D4SU00267A
2. de Andrade I, Fernandes A et al (2025) Bio-based polyurethanes: a comprehensive review on advances in synthesis and functionalization. Polym Int 74(11):941–957. https://doi.org/10.1002/pi.70010
3. Solvent-free polyurethane coating 2025 trends and forecasts 2033: analyzing growth opportunities. https://www.marketreportanalytics.com/reports/solvent-free-polyurethane-coating-179123. Accessed 18 Aug 2025
4. Xiao Y, Fu X, Zhang Y, Liu Z, Jiang L, Lei J (2016) Preparation of waterborne polyurethanes based on the organic solvent-free process. Green Chem 18(2):412–416. https://doi.org/10.1039/C5GC01197C
5. Desai Y, Jariwala S, Gupta RK (2023) Bio-based polyurethanes and their applications. In: Polyurethanes: preparation, properties, and applications, Advanced applications, vol 2. American Chemical Society, pp 1–14. https://doi.org/10.1021/bk-2023-1453.ch001
6. Wray HE, Luzzi S, D'Arrigo P, Griffini G (2023) Life cycle environmental impact considerations in the design of novel biobased polyurethane coatings. ACS Sustain Chem Eng 11(21):8065–8074. https://doi.org/10.1021/acssuschemeng.3c00619
7. Ling Z, Gu L, Liu S, Su Y, Zhou Q (2025) Non-isocyanate polyurethane from bio-based feedstocks and their interface applications. RSC Appl Interfaces 2:1123–1142. https://doi.org/10.1039/D5LF00160A
8. Habets T, Grignard B, Detrembleur C (2025) Non-isocyanate polyurethanes at room temperature—a dream becoming reality. Prog Polym Sci 165:101968. https://doi.org/10.1016/j.progpolymsci.2025.101968
9. Orabona F et al (2025) Cutting-edge development of non-isocyanate polyurethane (NIPU) foams: from sustainable precursors to environmental impact evaluation. Green Chem 27(25):7403–7444. https://doi.org/10.1039/D4GC05796A
10. Zeng Y, Li H, Li J, Yang J, Chen Z (2023) Preparation and characterization of solvent-free anti-corrosion polyurethane-urea coatings. Surf Interfaces 36:102504. https://doi.org/10.1016/j.surfin.2022.102504

11. Luo H et al (2024) Solvent-free polyurethane coatings based on an in-situ curing strategy at room temperature with self-cleaning, anti-corrosion and anti-graffiti properties. Prog Org Coat 188:108261. https://doi.org/10.1016/j.porgcoat.2024.108261

12. Solvent free UV curable waterborne polyurethane acrylate coatings with enhanced hydrophobicity induced by a semi interpenetrating polymer network. Scientific Reports. https://www.nature.com/articles/s41598-025-04739-1. Accessed 18 Aug 2025

13. Vogt-Birnbrich B (1996) Novel synthesis of low VOC polymeric dispersions and their application in waterborne coatings. Prog Org Coat 29(1):31–38. https://doi.org/10.1016/S0300-9440(96)00627-3

14. Aramayo MAF et al (2024) Eco-friendly waterborne polyurethane coating modified with ethylenediamine-functionalized graphene oxide for enhanced anticorrosion performance. Molecules 29(17):4163. https://doi.org/10.3390/molecules29174163

15. Tang S, Lei B, Feng Z, Guo H, Zhang P, Meng G (2023) Progress in the graphene oxide-based composite coatings for anticorrosion of metal materials. Coatings 13(6):1120. https://doi.org/10.3390/coatings13061120

16. Yuan X, Wang W, Du C, Kang Q, Mao Z, Chen S (2024) A novel noise-reducing and anti-corrosion polyurethane elastomer coating material modified by MXene/porous TiO2. Surf Interfaces 48:104256. https://doi.org/10.1016/j.surfin.2024.104256

17. Umayanga UGE (2024) Bio-based polyurethane coatings: pioneering efficient and eco-friendly solutions for sustainable applications

18. Smit AT et al (2023) Tuning the properties of biobased PU coatings via selective lignin fractionation and partial depolymerization. ACS Sustain Chem Eng 11(18):7193–7202. https://doi.org/10.1021/acssuschemeng.3c00889

19. Delavarde A et al (2024) Sustainable polyurethanes: toward new cutting-edge opportunities. Prog Polym Sci 151:101805. https://doi.org/10.1016/j.progpolymsci.2024.101805

20. Circularity of polyurethanes. Polyurethanes. https://www.polyurethanes.org/sustainability/circularity-of-polyurethanes/. Accessed 19 Aug 2025

21. Vitrimer synthesis from recycled polyurethane gylcolysate. ResearchGate. https://www.researchgate.net/publication/372608826_Vitrimer_synthesis_from_recycled_polyurethane_gylcolysate. Accessed 19 Aug 2025

22. Zhou W et al (2023) Biobased aliphatic polyurethane Vitrimer with superior mechanical performance and fluorescence-based defect diagnostic function. ACS Appl Polym Mater 5(4):3129–3137. https://doi.org/10.1021/acsapm.3c00272

23. Recyclable vitrimer-based printed circuit board for circular electronics. ResearchGate. https://www.researchgate.net/publication/373363907_Recyclable_vitrimer-based_printed_circuit_board_for_circular_electronics. Accessed 19 Aug 2025

24. Rossignolo G, Malucelli G, Lorenzetti A (2024) Recycling of polyurethanes: where we are and where we are going. Green Chem 26(3):1132–1152. https://doi.org/10.1039/D3GC02091F

25. O'Dea RM, Nandi M, Kroll G, Arnold JR, Korley LTJ, Epps THI (2024) Toward circular recycling of polyurethanes: depolymerization and recovery of isocyanates. JACS Au 4(4):1471–1479. https://doi.org/10.1021/jacsau.4c00013

26. Patti A, Acierno D (2023) Structure-property relationships of waterborne polyurethane (WPU) in aqueous formulations. J Vinyl Addit Technol 29(4):589–606. https://doi.org/10.1002/vnl.21981

27. Abrahamsen GM et al (2025) Morphology control in waterborne polyurethane dispersion nanocomposites through tailored structure, formulation, and processing. Langmuir 41(16):10383–10393. https://doi.org/10.1021/acs.langmuir.5c00226

28. Polyurethane coatings: a perfect product class for the design of modern automotive clearcoats. ResearchGate. https://www.researchgate.net/publication/329868246_Polyurethane_coatings_a_perfect_product_class_for_the_design_of_modern_automotive_clearcoats. Accessed 19 Aug 2025

29. Karlsson D, Dahlin J, Skarping G, Dalene M (2002) Determination of isocyanates, aminoisocyanates and amines in air formed during the thermal degradation of polyurethane. J Environ Monit 4(2):216–222. https://doi.org/10.1039/B110593K
30. Daily Planet. Safety tips for working with polyurethane. https://dailyplanetdc.com/2022/07/27/safety-tips-for-working-with-polyurethane/. Accessed 19 Aug 2025
31. Amado JCQ (2019) Thermal resistance properties of polyurethanes and its composites. In: Thermosoftening plastics. IntechOpen. https://doi.org/10.5772/intechopen.87039
32. Niedźwiedź MJ, Demirci G, Kantor-Malujdy N, El Fray M (2023) Influence of photoinitiator type and curing conditions on the photocuring of soft polymer network. Materials 16(23):7348. https://doi.org/10.3390/ma16237348
33. Li S et al (2025) Oxygen inhibition suppression by photoinduced electron transfer in oxime ester/triarylalkylborate photoinitiators. Macromolecules 58(4):2053–2064. https://doi.org/10.1021/acs.macromol.4c02565
34. Study on press formability and properties of UV-curable polyurethane acrylate coatings with different reactive diluents. https://www.mdpi.com/2073-4360/15/4/880. Accessed 19 Aug 2025
35. Safety considerations using urethane products. https://crosslinktech.com/support/health-and-safety/safety-considerations-using-urethane-products.html. Accessed 19 Aug 2025
36. Burke L. 7 safety measures for polyurethane spray foam installation. https://www.polymac-usa.com/7-safety-measures-for-polyurethane-spray-foam-installation/. Accessed 19 Aug 2025
37. Huuskonen P et al (2023) Occupational exposure and health impact assessment of diisocyanates in Finland. Toxics 11(3):229. https://doi.org/10.3390/toxics11030229
38. Jogie JA (2025) A comprehensive review of occupational medicine in the paint industry: roles of family medicine and internal medicine. Cureus 17(1):e76970. https://doi.org/10.7759/cureus.76970
39. Reeb-Whitaker C et al (2012) Airborne isocyanate exposures in the collision repair industry and a comparison to occupational exposure limits. J Occup Environ Hyg 9(5):329–339. https://doi.org/10.1080/15459624.2012.672871
40. Bello D et al (2007) Skin exposure to isocyanates: reasons for concern. Environ Health Perspect 115(3):328–335. https://doi.org/10.1289/ehp.9557
41. Airborne isocyanate exposures in the collision repair industry and a comparison to occupational exposure limits. https://www.google.com/search?q=Airborne+Isocyanate+Exposures+in+the+Collision+Repair+Industry+and+a+Comparison+to+Occupational+Exposure+Limits&rlz=1C1FHFK_en-GBPK1170PK1170&oq=Airborne+Isocyanate+Exposures+in+the+Collision+Repair+Industry+and+a+Comparison+to+Occupational+Exposure+Limits&gs_lcrp=EgZjaHJvbWUqBggAEEUYOzIGCAAQRRg7MgYIARBFGD3SAQc4MjBqMGo3qAIAsAIA&sourceid=chrome&ie=UTF-8. Accessed 19 Aug 2025
42. Testing of glove efficacy against sprayed isocyanate coatings utilizing a reciprocating permeation panel. https://www.google.com/search?q=Testing+of+glove+efficacy+against+sprayed+isocyanate+coatings+utilizing+a+reciprocating+permeation+panel&rlz=1C1FHFK_en-GBPK1170PK1170&oq=Testing+of+glove+efficacy+against+sprayed+isocyanate+coatings+utilizing+a+reciprocating+permeation+panel&gs_lcrp=EgZjaHJvbWUqBggAEEUYOzIGCAAQRRg7MgYIARBFGD3SAQc4MThqMGo3qAIAsAIA&sourceid=chrome&ie=UTF-8. Accessed 19 Aug 2025
43. Mellette MP, Bello D, Xue Y, Yost M, Bello A, Woskie S (2019) Evaluation of disposable protective garments against isocyanate permeation and penetration from polyurethane anticorrosion coatings. Ann Work Expo Health 63(5):592–603. https://doi.org/10.1093/annweh/wxz032